Get the Job You Want in IT:

Insider Strategies for a Successful Job Search Campaign

Ian O'Sullivan   John McClure

Get the Job You Want in IT
Insider Strategies for a Successful Job Search Campaign
by Ian O'Sullivan and John McClure

Copyright © 2010 by Signalman Publishing

Signalman Publishing
3209 Stonehurst Cir.
Kissimmee, FL 34741
407.343.4853

Find us on the Web at www.signalmanpublishing.com
Also, visit this book's website at: www.itdreamcareer.com
For comments or inquiries, email info@signalmanpublishing.com

Cartoons contained herein are copyrighted by Mark Parisi and printed with permission.

NOTICE OF RIGHTS
All rights reserved. No part of this book may be reproduced or transmitted in any form or by any means, electronic, mechanical, photocopying, recording, or otherwise, without prior written permission of the publisher. For information on getting permission for reprints and excerpts, contact info@signalmanpublishing.com.

NOTICE OF LIABILITY
This publication is the author's opinion on all subject matter contained herein. It is intended to provide helpful and informative material on the subject matter covered. It is sold or otherwise provided with the understanding that the author and publisher are not engaged in rendering professional services. If the reader requires personal assistance or advice, a competent professional should be consulted.

The publisher and the author disclaim any personal liability, loss, or risk incurred as a result of the use and application, either directly or indirectly, of any advice, information, techniques, or methods presented herein.

ISBN: 978-0-9840614-7-1
LCCN: 2010929211

Dedicated to Gina O'Sullivan

"This book is dedicated to my wife, Gina, for all her patience, support, and inspiration. She reminds me daily of the important things in life."

In Memory of

Wade McClure

Faith and Discipline

TABLE OF CONTENTS

Chapter 1: Getting Started. 7
How to plan and execute your job search campaign.

Chapter 2: Map your Career Path. 17
Know your destination, have the end in mind.

Chapter 3: Establish your Professional Values. 25
Understand the skills and value that you bring to a potential employer.

Chapter 4: Identify your Specific Talents. 33
Understand the skills that matter the most, the skills that are your true talents, and the ones that are your strengths.

Chapter 5: Study your Career History. 41
Reflect on the current or past positions you have held.

Chapter 6: Become a Better Story Teller. 49
Understand the power of your stories and how to convey them with impact.

Chapter 7: Craft your Personal Brand. 61
Build your personal brand, remember, you are the message.

Chapter 8: Build an Effective Resume. 65
Create an effective resume—one that will get you in the door.

Chapter 9: Optimize Your Digital Search. 85
A successful job search requires a variety of tools, and using the internet is one of them.

Chapter 10: Develop Your Professional Network.. 95
Leverage your personal network and the multitude of web based services.

Chapter 11: Sharpen Your Interviewing Techniques. 105
Practice interviewing techniques that will put you on the short list.

TABLE OF CONTENTS (continued)

Chapter 12: Seal the Deal..................................... 117
Learn negotiating strategies to get the compensation you deserve.

Bibliography... 125

Chapter One

Getting Started

The credit belongs to the man who is actually in the arena, whose face is marred by dust and sweat and blood, who knows the great enthusiasms, the great devotions, and spends himself in a worthy cause; who at best, if he wins, knows the thrills of high achievement, and, if he fails, at least fails daring greatly, so that his place shall never be with those cold and timid souls who know neither victory nor defeat.
—John F. Kennedy

How many times at social events have you been asked, "So what is it that you do Bob?" Do you enjoy being asked that question, or do you loathe it? Maybe you don't care either way. Do you have a great answer, or is your answer, "I'm in IT"? Technology is such an exciting field that your response to that question should ooze with passion and energy. Obviously, you must enjoy your work to enjoy speaking about it. Some people stumble onto the perfect job, the rest of us need a little luck and some effort. This book outlines the necessary steps to take in finding that perfect job in a very systematic and easy-to-follow process.

Undoubtedly, you have heard the term "best practices". A term used as some "holy grail" of procedures to be sought after and adopted. A best practice is a technique, method, process, or activity believed to be more effective at delivering a particular outcome than any other technique making you and your organization the best at what you do; whether it be product development, software delivery, customer support, etc. Best practices can be defined as the most efficient and effective way of accomplishing an objective based on repeatable procedures that have proven themselves over time. The idea is that any activity has a set of best practices that, if followed, a desired outcome can be delivered with fewer problems and unforeseen complications. If you are to be successful with your job search, following a proven set of best practices is essential.

If you have never conducted an active job search campaign before, or have not done so in a long time, this guide is for you. It will lead you step-by-step in the conduct of your own personal Information Technology Professional job search campaign. The practices outlined in this guide are proven, and if followed, will make your search more effective and ultimately more likely to be successful than if you did not follow the system. A successful job search campaign will set the stage, potentially, for the next several years of your professional life. Perhaps even the rest of your professional life. That said, this campaign requires discipline and focus.

The authors of this book have between them over 40 years of hiring process experience. We have been both the hunters and the hunted for small, mid-size, and large U.S. corporations (including Fortune 500). As the hunted, technology leaders and hiring managers, we have reviewed thousands of resumes, conducted hundreds of phone interviews (both technical and non-technical), and have participated in countless face-to-face interviews including 1 on 1 interviewing and panel interviewing. Through this, we have seen the good, the bad, and the truly awful, as it were. We know through experience the things a candidate should and should not do in order to be successful. As the hunter (or job seeker), we have applied for positions from entry level IT all the way up to corporate officer and everything in between. If there is one thing we've learned, it's the fact that the higher you get on the corporate ladder and the more advanced your career is, the harder the job search and the more competitive the process becomes. If you study and adopt the techniques that are presented in this guide, you could be ahead of 98% of the job seekers out there. That means when you are competing against a dozen other equally qualified candidates for the same position, you will have the edge. You will have the edge because you will be more effective at demonstrating and communicating the value that you bring to your prospective employer.

This guide is meant to be a practical, step-by-step workbook in which you can print out the pages and follow and write your responses in the activities. This workbook can be completed by yourself in a stand-alone fashion. It can be even more effective, however, if it is done with the help of another person such as a spouse, good friend, parent, or career coach. The one who helps can be used as a sounding board for reviewing your material and commenting

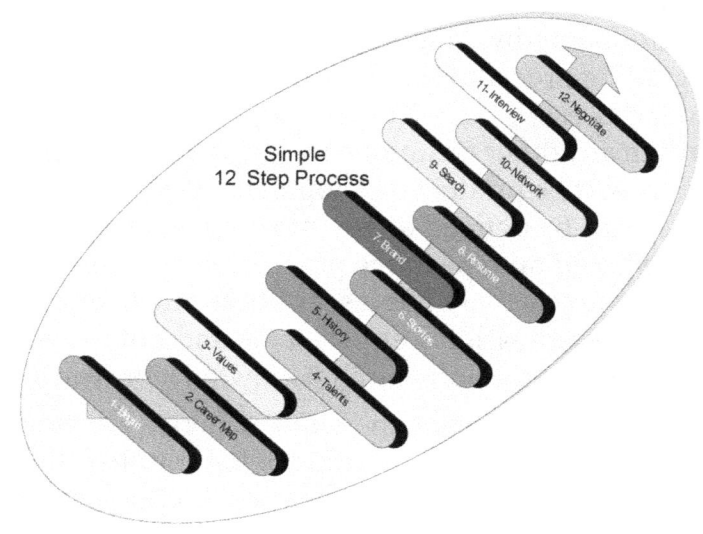

on what makes sense. This person can also be an important accountability partner—or one who helps to keep you on track and focused in your campaign. Even so, if you do not have a helper, don't worry. This workbook has the tools you need—you just need to be disciplined in reading the material, doing the activities, and following this twelve step process.

This simple process uses job search best practices that have been proven to work for people who understand and apply these to their campaign. This guide will walk you through the process by teaching you powerful techniques that will make your campaign highly effective. You have already accomplished Step one, which was to recognize that you need assistance with your campaign and as a result you have acquired this book. Congratulations! Now grab your No.2 pencil and let's work through the other eleven steps.

Step 2/Chapter 2: Map Your Career Path

You have made a decision to seek out your next professional opportunity. You have decided to move on from your current position or maybe that decision was made for you. Either way, you now have the opportunity to closely examine your career path. Certainly your next job will determine the amenities you will be able to enjoy such as the home you live in, the car you drive, the holiday trips you'll take, and the schools your children attend. But have you considered the effects the job will have on you as a person and ultimately the people around you, especially your spouse and children? I once found myself stuck in an eight-to-six job that was monotonous and boring, I did not feel challenged and had no particular interest in the work that I was doing. I snoozed the alarm eight times before dragging myself out of bed in the morning and when at work I completed my daily tasks like a robot. I impatiently watched the clock crawl toward the end of the day and as a result, my performance declined over time. If you do not spend time examining and re-examining your career path, you risk getting lost in a lifetime of dissatisfaction and unhappiness. It is completely up to you to manage your career. It is a lifelong process, and you will need to be proactive, resourceful, and self-aware. But it is simple. It is as simple as defining your short term goals, your long term goals and the intermediate steps and jobs you will need to take to achieve your dreams. This step will help you identify a career track that fits your skills and ambitions.

Step 3/Chapter 3: Establish Your Professional Values

In this step, you'll conduct a self-appraisal and think about what you want to accomplish. You will identify and record the values that are most important to you in terms of the job you do and the career you have. To do this you must first figure out who you are, and what you want and need in a job. We all have personal needs and we are all motivated by different things. For some, having a secure job is the most important factor. For others, having prestige is more important. And others don't really care what job they have as long as they're making plenty of money. A good friend of mine accepted a low paying position with very little opportunity for advancement so that he could be home every night for dinner with his family and coach his son's sports teams without worrying about work getting in the way. Another friend deals with a long commute and stressful job because of the opportunity and high pay. While we all look for something different in a job, there are some needs we all share, and that is to be treated with respect and integrity and to feel like we belong. After all, a person's greatest emotional need is to feel appreciated. Your personal and professional needs must be considered when planning your career because they should influence your choice of position and organization.

Step 4/Chapter 4: Identify Your Skills and Talents

Your skills and talents are vital to your future career success. What you like to do and what you are confident in doing play an important role in your future career satisfaction and success. You'll need to take inventory of your skills, talents, and strengths. Skills are abilities you develop, talents are born, and strengths make you the best at what you do. A skill is a capacity you have developed that can be used in different ways. It is said that by our late teens we have acquired more than 400 work focused skills. Can you imagine how many you have developed by your 30's, your 40's, your 50's? Unfortunately, most of us are using only a small portion of our skills and the majority of us are using the wrong ones in our daily activities. We are not using the skills that are our true talents and our true strengths. Finding your dream job requires that you understand the skills that matter the most, the skills that are your true talents, and the ones that are your strengths. This step will help you flush out your talents and identify your strengths.

Step 5/Chapter 5: Study Your Career History

I can't tell you how many times I stumped a candidate during a face to face interview by asking them the simple question, "Tell me about your position at ABC Corp," with ABC Corp listed on the resume as a previous employer. If you don't have your career history fresh and detailed in your mind you are not going to leave a good impression. In fact, you might even give the interviewer the impression that your resume is "not all that accurate". Every position you mention on your resume must be fresh in your mind, with the important aspects of the job ready to roll off your tongue, especially the ones that are pertinent to the job you are interviewing for. And if you have to look at your resume and read it in response to questions during the interview... you are doomed. Have your job history fresh in your mind, practice on the way to the interview without your resume in front of you. This step will help you refresh your memory so that you can speak about your prior work history with detail, accuracy and confidence.

Step 6/Chapter 6: Become a Better Story Teller

That's my story and I'm sticking to it. Human beings have been communicating with each other through storytelling since we lived in caves and sat around fires exchanging tales. You will be asked during an interview to tell a story. Actually, you'll be asked to tell many stories during the interview, for instance; "Tell me about a situation where you faced an unrealistic deadline during your time with XYZ Corp". Most in-person interviews today are geared towards what is known as "behavioral interviewing". The candidate is asked to describe a particular challenging situation (tell a story) from a previous job and explain how they handled it. It is based on the notion that what you did in the past is an indication of what you will do in the future in a similar situation. It is extremely important to get it straight. This step will help you build the stories that will capture your audience.

Step 7/Chapter 7: Craft Your Personal Brand

You are the product, and the product must be attractive to a potential employer. We all have an image, whether we are conscious of it or not. Ask yourself, does my image reflect my professional standards? When you walk into a meeting or an interview, does your appearance say who you are and does it demonstrate the image you desire others to perceive? Clothing and appearance are among the most important criteria we use to judge other people. Conservative business clothing covers 90% of your body; naturally it makes a very powerful statement. You can easily influence how someone else thinks of you by choosing the appropriate attire. After all, you never get a second chance to make a first impression. When you walk into that interview, you want to look successful and successful people generally dress successfully. The message you want to send is that you feel good about yourself and that you are confident. If you feel confident about your clothes, hairstyle and make-up you will present a positive and confident image. Fact: 55% of your impact on someone is based on your appearance, body language, and the way you take up space. 38% is based on how you speak (tone, volume, pitch and pace). And 7% is on the words you choose. Think about that, only 7% is based on what you say. In this step you will learn how to prepare the other 93%

so that you will have 100% working each time you encounter a potential employer during your campaign.

Step 8/Chapter 8: Build an Effective Resume

I once opened a position for a Senior Java developer, and was out of the office for a week after the job was posted. When I returned I had a stack of candidate resumes six inches high on my desk, and this was after the recruiter did a preliminary screening. When applying for a job, your resume will be one of many if not hundreds of others. So how do you get that hiring manager to notice yours? A large, colorful resume will stand out and it will surely be noticed, but is it effective? I once received a three page cover letter accompanied by a nine page resume. Another applicant tried to make an impression by using five different font types, three different font colors and a variety of highlights such as underlining and bolding. Needless to say, both these resumes made it straight into my trash. You cannot underestimate the importance of your resume. This step will help you construct a resume that will capture your potential employer's attention and serve as your ticket into the interview. Because without a good resume, you are not even going to get your foot in the door.

Step 9/Chapter 9: Optimize your Digital Search

The Internet is a virtual gold mine of employment resources. A successful job search requires a variety of tools and using the internet is one more way of improving your odds in finding that perfect opportunity. As part of your job search campaign, you must leverage the many job search engines the internet provides. A job search engine is a website that facilitates job hunting (also known as a job board). With most of these sites, you can upload your resume and submit an application to potential employers. These sites range from large scale generalist boards to specific job boards such as IT only jobs. Dice.com for example, is a job search engine dedicated to only finding technology jobs. It offers a targeted niche space for finding exactly the technology position you are looking for. This step will show you where to find these sites and how to make them work for you.

Step 10/Chapter 10: Develop your Professional Network

During your job search campaign, you'll need to refresh and extend your professional network. People and potential job opportunity contacts are everywhere. Try to meet as many people as you can during your campaign, collect business cards, introduce yourself at social functions and ask that question, "So what is it that you do Bob?"

Think about and prepare questions that will get the other person speaking about their jobs, especially people in the same field of work. Social networking web sites such as Facebook are now blooming in the business world, thanks to new networks that enable professionals to rub virtual elbows and share ideas and experiences. Millions of professionals already turn to broad-based networking sites such as LinkedIn to swap job details and contact information. These sites are often used for recruiting purposes. One of the most effective ways to land a job is through networking. The internet is buzzing with social media, this step will help you understand the many ways to use it in order to network, and eventually find that perfect job.

Step 11/Chapter 11: Sharpen your Interviewing Techniques

No matter how many degrees you have or how much experience; no matter what you know or who you know; if you aren't able to interview successfully, you won't get the job. The interview is the most important step in your job search campaign. The interview is a systematic, purposeful conversation. Your goal is to show the employer that you have the skills, background, and ability to do the job and that you can successfully fit into the company and be part of its culture. The interview can also be an opportunity for you to learn more about the position, the organization, and your potential supervisor allowing you to determine whether the job is right for you. Preparation, confidence, enthusiasm, and good communication skills are the key factors. This step will cover the basics and prepare you for a very successful interview.

Step 12/Chapter 12: Seal the Deal

Most of us feel we are underpaid for our hard work and dedication. Salary.com published the following statistics: 61 percent of people ages 18 to 34 and 55 percent of people ages 35 to 54 say they are not compensated fairly. The topic of money and salary will come up during your job interview. You must be prepared to properly handle the inevitable compensation discussion and negotiate for the salary you want and that you need. Think of it like bidding on a house, go to the negotiating table with three numbers in mind, a high, a low and a mid-point. The high in this case is your target and the low is the minimum you'll accept. Be prepared to give a range if asked, the range would be your mid-point to your high. For example, if your high is 115k and your low is 95k then your asking range should be 105k to 115k, with 95k being you absolute lowest. By having these numbers in your head in advance, you'll be prepared to have this discussion with confidence and ease. This final step in the process will prepare you for this fairly awkward, but extremely important conversation.

Helpful Resources

To assist you with this process, here are additional resources that are very helpful in understanding your goals, values, skills, and strengths. I have personally used these resources and have found that they bring a level of insight that helps inform my career choices and decisions.

The first is the *Now, Discover Your Strengths* series and the more recent iteration *Strengths Finder 2.0*. The premise behind the findings of authors Marcus Buckingham and Donald Clifton is that too many people spin their wheels trying to improve upon their weaknesses rather than playing to their strengths. They contend that we would be far more effective and successful if we only realized what our strengths are and then focus on building upon those. For example, let's say you have a talent around writing but are terrible at public speaking. Should you then go into a career such as acting which requires you to speak in front of large crowds? Or should you go into a career as a playwright where you are still involved in the theater that you love yet you are playing to your strengths as a writer?

Large organizations are starting to understand this concept and are having their employees undertake the Strengths Finder survey so that they can better understand and capitalize on their strengths. The website www.strengthsfinder.com allows one to purchase the latest book and take the survey so that you can better understand your strengths which may help you make some important career choices. I have found that the concept itself is the most valuable aspect of the book—playing to one's strengths rather than trying to compensate for one's weaknesses.

Another tried and true resource that has been adopted and used by companies and organizations for years is the Meyer's Briggs Type Indicator (MBTI). The premise behind the MBTI is that it measures an individual's personality type as preferences in four different areas:

1) Extroversion/Introversion – Measures the extent to which you focus on either the outer world or your own inner world. Another way to think of this is where you get your energy. Do you get it from others (extroversion), or do you get it internally (introversion).

2) Sensing/Intuition – Determines how you process information. Are you more likely to focus on the information that you take in (sensing), or do you add interpretation and meaning to that information (intuition).

3) Thinking/Feeling – Measures the process you use to make decisions. Are you more inclined to look first at logic and consistency (thinking), or are you more likely to first consider the people and circumstances (feeling).

4) Judging/Perceiving – Measures your orientation to the outer world and how others see you. If your preference is "judging" you will appear more task oriented and planned. If your preference is "perceiving" you will appear more spontaneous and flexible.

Understanding the combination of the 4 preferences in each of these areas results in a matrix of 16 personality types. Through years and years of processing these surveys, and then following up to determine which personality types tend to gravitate towards certain occupations, the folks at Meyers Briggs have published some interesting findings and correlations showing that indeed certain personality types are more prevalent in certain professions. You can get more information and find out how you can take the survey by going to their website: www.meyersbriggs.org .

To those who are skeptical of these tools, I would just add that knowledge is power and the more understanding and education you have about yourself, the more insight you will have to make better decisions when it concerns your career.

"When I was 17, I read a quote that went something like: 'if you live each day as if it was your last, someday you'll most certainly be right.' It made an impression on me, and since then, for the past 33 years, I have looked in the mirror every morning and asked myself: 'If today were the last day of my life, would I want to do what I am about to do today?' And whenever the answer has been 'No' for too many days in a row, I know I need to change something."

——Steve Jobs

Chapter Two

Map Your Career Path

The best way to predict the future is to invent it.
—Alan Kay

I want to put a ding in the universe.
—Steve Jobs

If you are like me, you started in this field as a technical resource. You have a very specific skill base such as software development, database administration, or network engineering. Or maybe you are not particularly technical and entered the IT field from the business side and brought thought leadership, product understanding and strategic thinking to the IT table. However you got here, ask yourself this: "what is my desired end goal?" The overwhelming majority of us begin our job hunt without first figuring out what it is that we really want to do. It's like heading to the airport, suitcase packed with no plane ticket and no particular destination in mind. Sound crazy? That's exactly what most job seekers do. They begin their job search without first deciding where they want to go.

In Stephen Covey's *The 7 Habits of Highly Effective People*, the author presents Habit #2: The ability to envision in your mind what you cannot at present see with your eyes. This notion is based on the principle that everything is created twice. First, there is the mental creation which can then be followed by the physical creation. The point is, you and only you must visualize what you want in life and set a course to achieve it, otherwise you are empowering others to do so. "Begin with the end in mind" means to begin each day, task, or project with a clear vision of your desired direction and destination. Your job search campaign and your career are no exceptions to this rule.

Do you want to stay technical and follow a programming or architectural path ultimately achieving the position of Chief Technology Officer (CTO)? (A position that allows you to remain in the trenches while playing a significant role in influencing the technology direction of the organization). Or do you want to follow in more of a leadership role, getting to know the business better and slowly moving away from the raw technology and eventually becoming the Chief Information Officer (CIO)? (A position where you will surround yourself with technical resources to handle the day-to-day aspects of the technology while you make strategic decisions that will impact the health and success of the company). You must decide what your long term ambition is, you must start with "the end in mind". With the end in mind, you will be ready to make the next big decision of your professional career.

The worksheet in this step will help you define your career path. The exercise is simple. You are to choose your career path with the end in mind. I do realize that every organization is different in terms of positions, titles, and even career paths. For example, I have mentioned the positions of CIO and CTO as top IT corporate positions. However, during your job search you might see titles such Vice President of Technology Delivery or Senior Vice President of Information Technology. You might see titles like Solutions Architect and Product Architect, where one requires solid technical abilities while the other asks for business product understanding with little or no hands-on technical skills. The point is, despite how organizations handle their IT shops, the skills needed to follow a particular career track stay the same. With that in mind, read through the eight career tracks presented and find *your* place in IT.

Entry Level	Junior Level	Senior Level	Supervisor Level	Manager Level	Director Level	VP Level	SVP Level
IT TRACK 1		Experience					
IT TRACK 2			Seasoned	Leader			
IT TRACK 3		Experience					
IT TRACK 4					Leader of Leaders	Strategy	Vision
IT TRACK 5		Experience		Leader			
IT TRACK 6			Seasoned				
IT TRACK 7		Experience					
IT TRACK 8							

Diagram 2.1

Software Engineering \ Application Development (Track 1)

The Software Engineering (or software development) career track calls for skilled professionals focused on the design and creation of software. Resources in the area are not necessarily just programmers, even though the foundational craft is the ability to read and write code. Resources in this area must have the ability to interact with both business functions and technical functions as they are often required to explain business needs to developers and technology restraints to non-technical business partners. This career track includes job titles such as software engineer, application developer, software architect, application programmer, web designer\developer, technical lead, technical project manager, solutions architect, technical director and Director/VP of delivery. The software engineering area of an IT shop typically accounts for the majority of the total resources, and therefore offers the most opportunity.

Information Management \ Database Administration (Track 2)

The Information Management path involves activities using database (DB) software to store and manage information. Responsibilities in this area include developing SQL (Structured Query Language) procedures, constructing data warehouses, performance tuning of data-stores, and ensuring data systems operate efficiently. Information Management is responsible for ensuring that the corporate data is stored effectively, backed up regularly, and secure from unauthorized access. This career track includes job titles such as data architect, database administrator (DBA), DB analysts, DB modeler, DB warehouse (DW) architect, DB developer, decision support systems analyst, business intelligence (BI) engineer, DW/BI management specialist, DW/BI project manager, BI architect, and VP of DW/BI. The typical craft for someone following this career path is the ability to write SQL. Information Management professionals often move into strategic leadership roles within the company such as Director of Information Management and eventually on to a Chief Technology Officer role.

Product Management (Track 3)

Depending on the company size and history, the positions within Product Management have a variety of functions and roles. The product management function is the hub of the many activities surrounding a particular business product. Professionals in this area lean more towards the business side than the technology side. This career track includes job titles such as business analyst, product manager, scope manager, Director of product management, Director of business technology consulting, and VP of business systems. In some companies the product team reports to the business side leadership while in others, this group reports to IT leadership.

Project Management (Track 4)

The Project Management track includes professionals that coordinate and report on resources, tasks, and financials associated with a technology initiative. This includes managing software development projects, networking projects, IT installations or conversions, or any other effort where business and technology needs have to be managed and resources have to be coordinated. This career path includes job titles such as project manager, project leader, project coordinator and program manager. There are various levels of project management and differences between the size and scope of projects to manage. Project managers can come from the technical ranks as well as the business ranks. A good project manager will often be promoted to a senior level role within IT, such as a Director of project management office (PMO) or VP of project management and software delivery. This career track includes resources with credentials such as PMI\PMP certification, Six Sigma discipline and black belt/master black belt certifications.

Network Engineering \ Administration (Track 5)

The Network Engineer track includes activities involving the design and maintenance of both the hardware and software necessary for computer networks. Professionals in network engineering are typically high level technical analysts with a specialty in Local Area Networks (LANs) or Wide Area Networks (WANs). Resources in this area will have a solid understanding of network apparatus, network operating systems, and networking protocols. This career track includes job titles such as network administrator, LAN administrator, network operations analyst, information systems administrator, network engineer, network technician, WAN manager, system architect, systems manager, and Director/VP of network administration. The typical craft for someone following this career track is the ability to write scripting languages. Many of the top paying jobs within the IT field are in Network Engineering.

Compliance \ Quality and Information Security (Track 6)

There are several different roles and resource types within the compliance group and career track. There is quality assurance, software testing, change management, release management and information security. This career track includes job titles such as security specialist, information security engineer, security administrator, QA tester, QA Manager, security architect, change advisory board manager, deployment\release manager, IS analyst, Manager of Application Assurance, VP/Director of Information security (or Compliance) and Chief Security Officer. Given increasing government regulations and audit requirements, the compliance space is one of the fastest growing parts of IT.

Technology Support \ Service Management (Track 7)

The Technology Support track includes resources that support internal and external customers. Professionals on this track have a firm understanding of one or more applications and systems, depending on their area of expertise. They focus mainly on aiding users, both internally as systems support and externally as customer support.

These resources are usually technical with a firm understanding of computer hardware and the inner workings of the components. Because of the interactive nature of this position, good interpersonal skills and communication skills are essential. This career track includes job titles such as desktop support analyst, PC specialist, help desk support analyst, service desk technician, field technician, desktop support manager, help desk manager, Director of customer services, and VP of technology service management. The activities in this IT area are typically driven by the company's trouble ticketing or incident management system. Resources are often educated in industry best practice processes such as ITIL. Information Technology Infrastructure Library (ITIL) has become a widely accepted approach to IT service management.

Technology Specialty (Track 8)

The Technology Specialist is a catch-all track and includes resources with specific skills and certifications. IT specialists can take on many roles such as telecommunications systems analyst, wireless specialist, back office applications specialist, IT trainer, technical writer, and IT generalist. IT Specialists typically have a degree in Computer Science, Information Science, or Information Systems Management. Select this career track if you do not fit into any of the previous ones. The other seven tracks cover the majority of IT positions but do not account for every possible variation and/or title.

Now that you are familiar with the different career tracks in information technology, you should be able to choose the track that suits your experience and skills, and the one that will take you to your desired career destination. To complete Step 2, use the IT organizational chart and choose your career track by identifying where you are currently (or last held position), your desired next position (short term goal) and finally your end point (long term goal). While you complete this step think about a quote by Zig Ziglar "*A goal properly set is, half reached*".

Refer to your chosen career track often during your campaign to ensure that you are staying on its intended course and are not wasting time pursuing a professional opportunity which may not be a stop along the way.

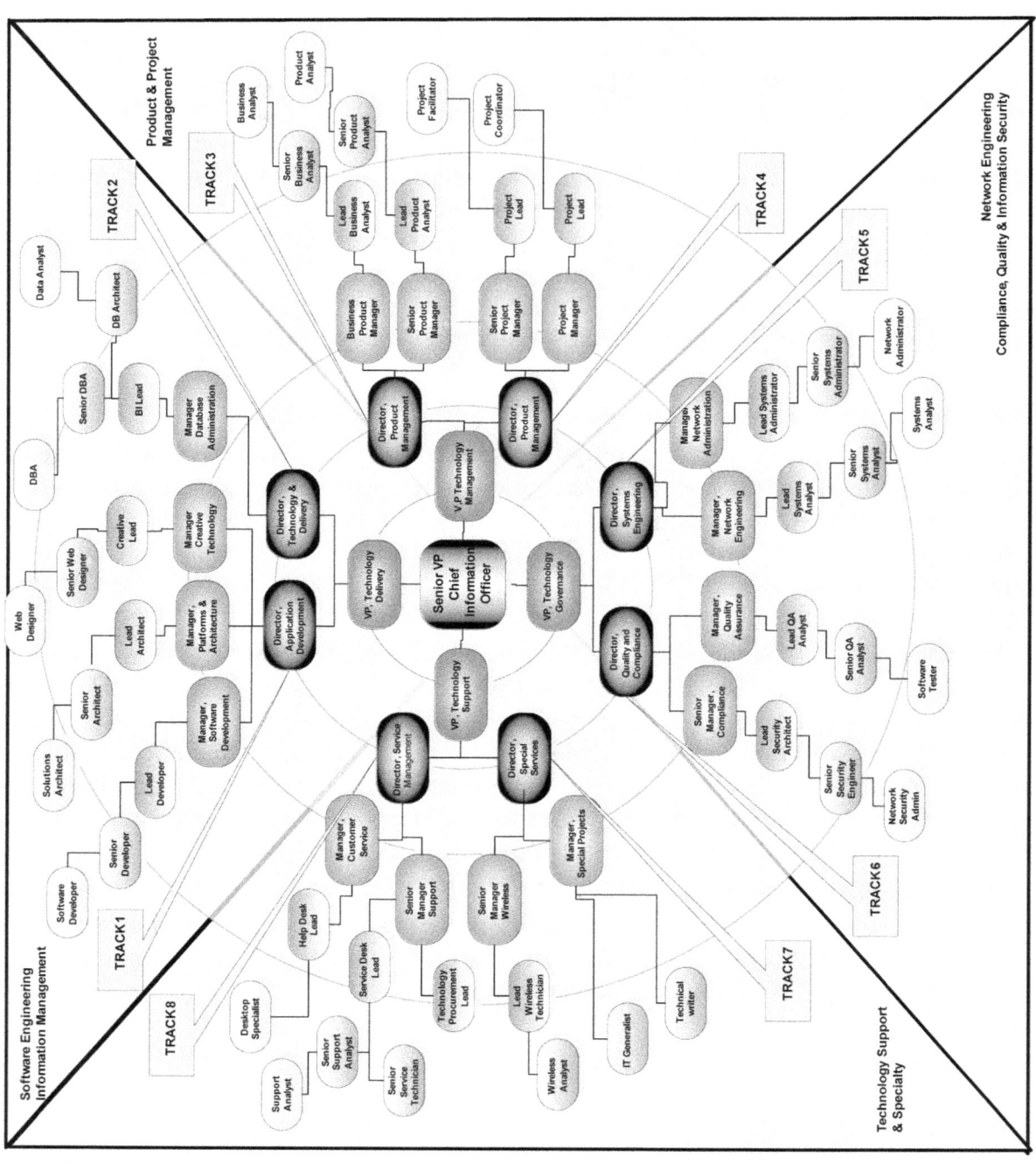

Diagram 2.2

Chapter 2: Map Your Career Path

	Current Position	Next Position	Destination
Track 0	Lead Developer	Manager, Software Delivery	VP, Technology Delivery
Track 1			
Track 2			
Track 3			
Track 4			
Track 5			
Track 6			
Track 7			
Track 8			

Track "0" shows an example entry. This and all the other workbook forms in this book can be found on this book's website: www.itdreamcareer.com.

Chapter Three

Establish Your Professional Values

Success is getting what you want. Happiness is wanting what you get.
—Dale Carnegie

What is success? What images come to mind when you think of the word and the state of success? I'll bet your first thoughts are that of material objects such as a fat bank account, a spectacular home or multiple homes, brand new cars, expensive clothes, traveling to distant places and staying at five-star hotels. And you know what, there is absolutely nothing wrong with wanting those things. But success has to mean more, it has to leave you feeling content and satisfied inside.

When Joe Torre took over as manager of the New York Yankees in 1996, the most storied franchise in sports had not won a World Series title in eighteen years. When Torre left New York eleven years later, he had four World Title rings and was considered one of the most beloved and successful managers in the game. In his book, *The Yankee Years*, Coach Torre talks about what success means to him. Oddly enough he doesn't speak of his multi-million dollar salary or any of the many World Series games he won. "To me," says Torre, "success is getting the most out of your ability. It's being able to go past that wall where you feel tired and frustrated. And once you go beyond that wall I think there's a great deal of pride involved. Life isn't easy and if you just sit down when you get tired instead of pushing on, then I don't think you ever realize satisfaction and self-esteem."

So what is success for you? Far too often we let others define success for us. We find ourselves reaching for that big house and expensive car just to impress others. Look inside yourself, and discover what feels good within. And always remind yourself that others may not share your definition of success, and that's okay because you are living your own life. Be true to yourself and follow your own path to your own definition of success.

Thinking about what success means will help flush out the values most important to you. Values significantly define who you are whether you are consciously aware of them or not. Every individual has a core set of personal values ranging from the simple (such as the belief in hard work), to the more psychological (such as self-reliance and concern for others). These values will greatly influence the decisions you make; they will inform and guide you through your personal life.

And then there are your professional values. These values are entirely subjective and may often differ from those of your spouse, parents, or best friends. Before you begin your job search campaign, it is extremely important to establish what you value and then look for an opportunity that will satisfy those values. If you don't, you may find yourself in the wrong position and you will be reminded on a daily basis that your work does not align with your professional values. Having a job and getting paid on a regular basis is a good thing, being miserable working in a position that misaligns with your values is not.

The worksheet in this chapter will help you work through your own understanding of your values. Read through the three lists and then write down the values that are most important to you and the ones that are least important to you. The goal is to create a list of the top three work "environment" values, the top three work "focus" values, and the top three work "reward" values. You will establish the job values that you cannot live without verses the ones you can live without. Start by using the process of elimination i.e. eliminate the values you do not feel a compelling need to have in your next position. The first three (from each list) that you eliminate immediately will become the values you can live without. Keep eliminating items until you have three left. Then rank these remaining values in order of highest priority first. The top three items on each list become your ***must have professional values***. The bottom three are the ones you can live without and the other six are the "*take 'em or leave 'em*" professional values.

Before you get started with this step, take a few minutes to reflect on past job activities. Think about the activities that have left you feeling good about yourself and the job activities that left you drained and unhappy. Remember to refer to your response here often to ensure that you are staying on track and that you are not wasting your time pursuing professional opportunities that may be in opposition to your professional values.

Work Environment (company culture)

1) **Continuous Learning:** An environment that presents opportunity for continued professional development, growth, and learning.

2) **Social Development:** An environment where you have opportunity to develop friendships with work associates.

3) **Consistency & Stability:** An environment where your activities and responsibilities

are consistent and not likely to change over short periods of time. Routine activities with predictable work projects.

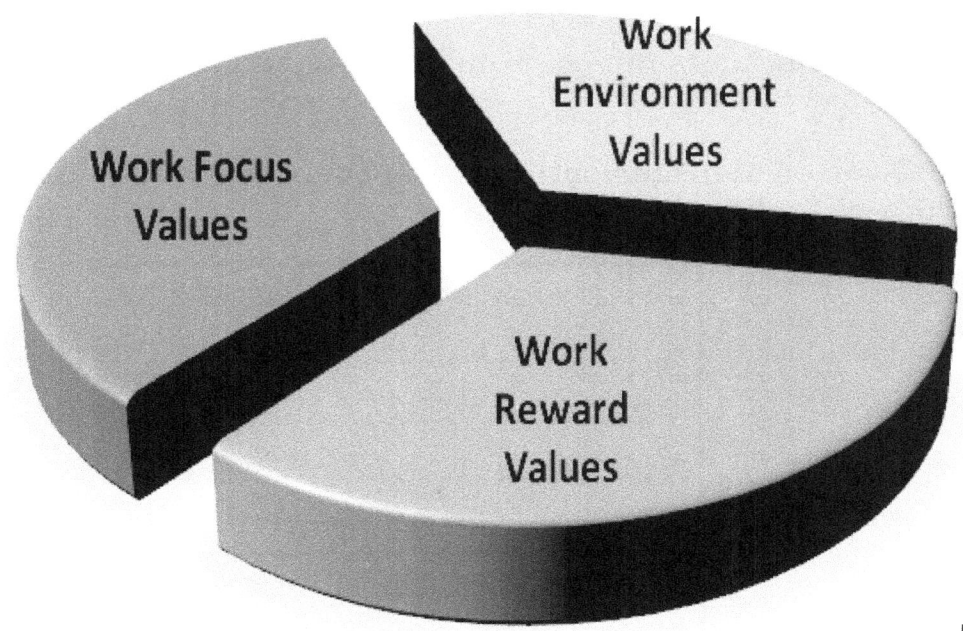

Figure 3.1
Professional Values.
3 parts make the whole.

4) Competitive Work: An environment where your skills and abilities are constantly compared to others, and compensated as such.

5) Working within Teams: Having a close working relationship with others as part of a team or group.

6) Change and Variety: An environment where your responsibilities will evolve in content and setting, providing variety and assortment of tasks and responsibilities.

7) Adventure and Innovation: A pioneering environment using cutting edge technologies and/or techniques filled with excitement and risk taking.

8) Flexible Schedule: Flexibility in commitments and schedule to allow for balance between work life and personal life. A position with a work schedule that allows you to pursue personal activities.

9) Working Location: Working in the geographic area of your choice, and/or working from your home office.

10) Working Travel: A job that requires a frequent need to travel.

11) Sound Leadership: An environment with strong leadership. Leaders act as role models and mentors.

12) Evolved Management: An environment with frequent guidance and clearly stated expectations from the leadership.

```
Three most important Work Environment Values
1. _____
2. _____
3. _____
```

```
Three least important Work Environment Values
1. _____
2. _____
3. _____
```

Work Focus (daily activity)

1) Helping Others: Contributing to societal good by helping others improve their lives.

2) People Connecting: Day-to-day contact with the public and/or people outside the organization.

3) Decision Making: Responsibility and accountability around making decisions.

4) Working Independence: Working without significant direction from others.

Autonomy and independence. Freedom with little or no supervision.

5) Leading and Directing: Leading a team and being directly responsible for the contribution made by others.

6) Creative Freedom: Leadership encourages and rewards creativity and imagination with a push to be innovative.

7) Accuracy and Precision: Focus on perfection with little tolerance for error.

8) Pressure Challenged: Focus on hard deadlines and time demands where resources are constantly under pressure.

9) Artistic & Aesthetical: Leadership encourages and rewards artistic thinking and behavior.

10) Intellectually Challenged: Constantly challenged and required to frequently solve problems.

11) Just in Time: A fast paced delivery with expectations that work gets done rapidly.

12) Less Bureaucracy: Open standards. Flexibility in process, methodology and tool selections.

Three **most** important Work Focus Values

1. _____
2. _____
3. _____

Three **least** important Work Focus Values

1. _____
2. _____
3. _____

Work Reward (compensation & benefits)

1) Financial Success: Competitive base salary and rewards-based compensation.

2) Job Security: A reasonably high probability that you will keep your job despite the economic climate.

3) Rewards and Recognition: Frequently rewarded for loyalty, dependability, and recognized for quality of work. Recognition for a job well done.

4) Intellectual Status: Being a recognized subject matter expert among your fellow workers and throughout the organization.

5) Social Status: Having prestige, notoriety, social status for your expertise among industry peers and throughout the technology industry.

6) Title Status: Having a position and title that is considered highly ranked and important.

7) Morally Fulfilling: Feeling that your work is contributing to a set of moral standards, that you feel are very important.

8) Promotion & Advancement: Clearly define advancement paths and opportunities for promotion.

9) Work Enjoyment: A job that gives you pleasure, where you have self-respect and pride in the work that you do.

10) Health Benefits: Company provided health insurance that pays for medical and dental expenses.

11) Fringe Benefits: Various non-wage compensations provided in addition to normal compensation (i.e. discounted hotel rooms when working for a hotel company).

12) Early Retirement: A position with an organization that has unusually high compensation scales and/or retirement investment plans that will enable you to retire at a younger age than may otherwise be achievable.

Chapter 3: Establish Your Professional Values

Three most important Work Reward Values

1. _____
2. _____
3. _____

Three least important Work Reward Values

1. _____
2. _____
3. _____

Chapter Four

IDENTIFY YOUR SPECIFIC TALENTS

The person who knows 'how' will always have a job. The person who knows 'why' will always be his boss.
—Diane Ravitch

Over the course of your professional career, you have acquired many skills either through work experience or educational endeavors. Being able to properly identify and communicate these skills is an important step towards getting the right job. The information captured in this step will be necessary first when contacting people in your network because you will then have the opportunity to speak about your skills, and secondly during interviews where you will be asked to describe your particular skills in detail. In this step, you will identify the skills that are your talents and strengths. Your talents are those special natural abilities and/or aptitudes you possess and are a subset of your skills; and your strengths are that special subset of your talents.

The results from this step will serve as input to step 5 where you will specify the job history that highlights your talents & strengths. And again in step 6 where you will build stories and learn how best to communicate these talents and strengths, hence convincing the decision makers that you have what they need.

Your activity in this section is to identify and document your abilities in a simple three step process. First, you will identify your professional skills—the things you are good at, the things you do well. Next, you will identify which skills you are proficient at—the skills that come natural to you, and the ones you feel are your true talents. Finally, you will pick only the skills that are required when performing activities that you enjoy. Just because you are good at something does not mean it is something you enjoy doing. That is why it is important to identify what skills you have and which of these skills fulfill you, energize you, and make you feel good when you are performing tasks that require them. This final list of five particular skills will be limited to the ones that are your best abilities and more importantly, the skills that you feel are your strengths.

In the introduction to this book, I mention a helpful resource and one of my personal favorite books, *Now, Discover Your Strengths*, by author Marcus Buckingham.

In his book, Buckingham discusses the importance of discovering your strengths and points out the incredibly positive results you can expect when you put them to work for you. *"Stop waiting for the world to notice how special you are,"* he writes, *"or waiting for others to recognize what you do well and push you at it, because they won't. It takes YOU to take the initiative to put yourself in a place where you can be challenged the way you need to be, a place where you have a chance to make a difference"*. His point is that no one knows you better than you, so spend the time identifying what you do best and find an opportunity where you can put your strengths to work.

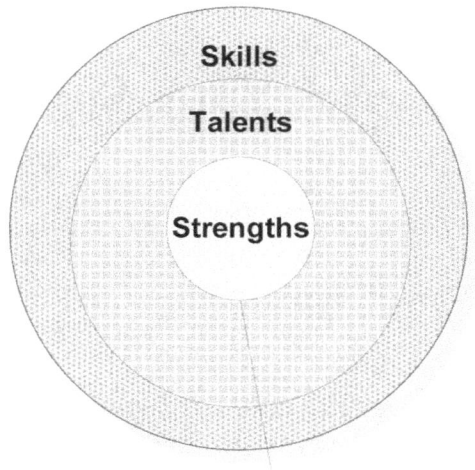

Diagram 4.1

The skills are presented in 3 lists: General Skills, Interpersonal Skills, and Technical Skills. Your task is to select 15 skills from the three lists. It's perfectly okay to pull most of your skills from one list or the other, there is no minimum or maximum you select from each list only that you should identify at least 15 skills to start. The Technical Skills list includes mostly hard skills, hands-on skills, which may not apply to you and that's fine, just ignore that list and select your skills from the other two lists. Undoubtedly, there are skills not listed that are particular to you and your experience. Feel free to include them, the point of the exercise to is capture the skills and in particular the strengths that are specific to you.

Once you have selected the 15 skills that you feel are a good match, then cut the list down to 10. This list of 10 will be the skills you are proficient at, the skills that come natural to you, the ones you feel are your true talents. Finally, cut the list down to five. Pick only the skills that are required when performing activities that you enjoy. These five particular skills will be the ones that are your best talents and more importantly, the ones that you feel are your strengths. These are the skills you will focus on when describing what you do best and what true value you bring to the table.

General Skills

Your general skills are the non-specific abilities that you have acquired or developed through training or experience. Identify and select your skills from this list and add them to the skills list in the workbook.

1) **Writing:** Communicating effectively through writing as appropriate for the needs of the audience.

2) **Speaking:** Talking to others to convey information effectively.

3) **Presentation:** Delivering a message in an effective manner to a particular audience.

4) **Decision making:** Defining action based on the information presented.

5) **Problem solving:** Evaluating facts, considering options, and implement solutions.

6) **Self-motivating:** Finding a reason and the necessary strength to do something.

7) **Self-discipline:** Directing yourself to take action regardless of your emotional state.

8) **Active learning:** Seeking information for problem-solving and decision-making.

9) **Critical thinking:** Using logic and reasoning to identify the strengths and weaknesses of alternative solutions, conclusions, or approaches to problems.

10) **Reading comprehension:** Understanding written sentences and paragraphs in books and documents.

11) **Financial management:** Determining how money will be spent to get work done, and accounting for these expenditures.

12) **Material resources management:** Directing and overseeing the appropriate use of hardware, facilities, and materials needed to do certain work.

13) **Time management:** Effectively managing one's own time and/or the time of others.

14) **Judgment:** Effectively considering the pros and cons of a decision and choosing the most appropriate action.

Interpersonal Skills (soft skills)

Your interpersonal skills are your ability to operate within business organizations through social communication and interactions. Identify and select your skills from this list and add them to the skills list in the workbook.

1) **Instructing:** Teaching others how to perform a certain task.

2) **Negotiating:** Effectively achieving agreement through discussion.

3) **Persuading:** Convincing others to change their minds or behavior.

4) **Communicating:** Conveying information through verbal and non verbal means.

5) **Coordinating:** Adjusting actions in relation to others' actions.

6) **Managing:** Coordinating and supervising others to accomplish an objective.

7) **Leading:** Influencing, guiding, and supporting others while they accomplish an objective.

8) **Motivating:** Driving others towards objectives and accomplishment of goals.

9) **Active listening:** Giving full attention to what other people are saying, taking time to understand the points being made, and not interrupting at inappropriate times.

10) **Monitoring:** Monitoring/assessing performance of yourself, other individuals,

or organizations to make improvements or take corrective action.

11) **Resource management:** Selecting, motivating, developing, and directing people as they work. Develop the capacities used to efficiently allocate resources.

12) **Team participation:** Collaborating with others using division of labor or other means in order to successfully accomplish the team's goals.

13) **Conflict resolution:** Intervening and negotiating in order to alleviate or eliminate sources of conflict. Bringing others together and trying to reconcile differences.

14) **Social perceptiveness:** Being aware of others' reactions and understanding why they react as they do.

15) **Service orientation:** Actively looking for ways to help people.

Technical Skills (Hard Skills)

These are your specific skills required of a specific job. These are the skills used to analyze, design, build, and operate the platforms and components of technological systems. As most of us have discovered, technical skill building is an on-going, never-ending process. Many Information Technology skills become obsolete relatively quickly.

LAN/WAN Management, Network Administrator
Unix Operating System Administration
Linux Operating System Administration
Windows Operating System Administration

Sun Java Language Coding
C++ Language Coding
Microsoft C# Language Coding
Perl Language Programming
SQL Database Management

XML - Extensible Markup Language
HTML Skills -Creative Design
Portal Technology Development
Firewall & Security

Wireless Technology
Infrastructure Management
Project Management
Product Management
Systems Analysis

Business Process Management
Content Management
Document Management
Knowledge Management

Database architecture and design
Unified Modeling Language (UML)
Use Case Modeling
Object Oriented Modeling

Software Development Methodologies such as the Rational Unified Process and the various Agile methodologies.

Usability analysis
Change Management
Application Assurance
Quality Assurance
Black box testing
White box testimg
Integration testing
Load testing
Automated testing (using various standard industry tool sets)

CHAPTER 4: IDENTIFY YOUR SPECIFIC TALENTS

Skills
1
2
3
4
5
6
7
8
9
10
11
12
13
14
15

Talents
1
2
3
4
5
6
7
8
8
10

Strengths
1
2
3
4
5

Chapter Five

Study Your Career History

A man who carries a cat by the tail learns something he can learn in no other way.
—Mark Twain

During the job application and job interview process you will be asked specific questions related to your current and previous positions. I can't tell you how many job applicants fumbled when I asked them simple questions about prior employment. You don't want to be one of these pretenders. This step will help you refresh your memory and record the specifics of the past positions you have held. If you are entry level, then you can skip this step. If your current or last position was your first in the IT industry, then this exercise will be simple, you'll have only one sheet to complete. If you are a seasoned veteran and have held several positions during your career, only list the last two or three positions, no more. Many years of relevant experience with a job history that charts your steady upward progression within IT is a very positive thing and should be listed on your resume. But for this exercise, don't go back more than 10 years, because although the positions you held more than a decade ago are important, they won't be discussed in detail during interviews.

The information captured in this step, detailed account of your work history, will be useful in several ways. It will provide the foundation for telling your stories in chapter 6, crafting your personal brand in chapter 7, and building your resume in chapter 8. It will also be necessary during the application process where you will be asked to list current and previous positions and during the interview process where you will be expected to review your work history in detail. One caution here, and I'm sure I don't have to say it, but I will anyway, "Always be honest with your accounts and descriptions," and try not to embellish. It's a small professional world out there where we are all connected. You don't know who the hiring manager will be reaching out to when checking your references and verifying your background. As a hiring manager, I always did my homework. Prior to an interview I would review the candidate's profile and ping my contacts who worked for the companies listed on the resume.

When Tonight Show host and popular comedian Jay Leno was interviewed for an article in Parade magazine, he was quoted as saying, *"It's not my nature to be frightened about starting a new job. I only get nervous around things I don't know or can't do well."* Leno went on to talk about how he prepares for life events and by doing so reduces any anxiety and nervousness. Granted, Leno's world is quite different than most of us in terms of audience size and compensation numbers. But like all successful people, Leno follows the fundamentals such as always preparing for big events. Now take this advice and apply it to your job search. Isn't this a big event in your life, one that requires thorough preparation? Try following the *3-P's* of public speaking approach: **Prepare**, **Practice** and **Present**. Prepare by completing the job history exercise in this chapter. Practice by reviewing the material often during your job search. And present by delivering the material when requested during the process. If you follow the *3-P's* approach you will find that when the time comes to present your job history you will be ready. And as a result, the material will feel that much easier to deliver and sound that much better to the listener. Be prepared, practice, and win over the audience when your opportunity to be center stage arrives.

The exercise in this step is simple. You will start by listing your employer's company name, the company's industry, the last position you held with them and the title of your supervisor. Listing the title of the person you reported to is important for this reason: it demonstrates your position within the company. Job titles are not always a good indication of the actual job itself. Job titles at different companies mean different things. When I consulted for a Wall Street financial firm, several senior level developers had the title of vice president even though their primary job was to write code. When I took a job for a national media company, I reported to the director who was essentially the CIO and he reported to the president of the division. The point is, your title doesn't always tell the whole story of what your job role was. Listing your *supervisor's title* will provide additional clarity and help explain your position. For instance, a manager reporting to a director suggests people and project responsibility verses a manager reporting to a vice president which suggests strategic and financial responsibility. Of course, you will have ample opportunity to discuss the specifics during the interview. However, getting to the interview requires you to be chosen based on what you specify on the job application or list on your resume. This is why it's critical to provide the right information, the pertinent information, when describing previous positions.

Next, provide a short description of what the company produced or provided. Spend a few moments to answer this one properly, and be prepared to discuss it. With the exception of technology companies, IT exists to serve the business. Show that you get this and that you understood the business of your previous employer.

If your previous position was a supervisory position then list the number of direct reports you had. Don't worry if the number is zero. Information Technology is full of individual contributors and people leadership is not a requirement for many positions. If you were responsible for assigning tasks to resources but did not manage them directly, then list the number of resources involved. A good example of this is the role of team lead. A team leader assigns tasks and provides guidance, but may not necessarily have Human Resource responsibility for the individuals performing the tasks. Achieving objectives by influencing and motivating others that are not directly reporting to you demonstrates a characteristic of a true leader. In the book, *You Don't Need a Title to Be a Leader* by Mark Sanborn, the author writes, "*As all leaders know, untitled or not, leadership is power with people, not power over people.*" This book is a must read for all professionals who want to take their leadership skills to a new, more effective level. Management and leadership are not the same thing, it is important to understand the distinction. Genuine leadership is not conferred by a title, or limited to the executive suite. Rather, it is shown through our everyday actions and the way we influence the lives of those around us.

List any financial responsibility you may have had in the form of dollar figures. This is especially important for higher level positions, director and above, where financial or P&L responsibility will be discussed in detail during interviews. Again, try not to exaggerate, the temptation is there to double and triple the figures. I interviewed a gentleman for a technical director position once and on his resume was the following bullet: *"Responsible for a $2M+ annual capital budget and accountable for a $25M approved technology capital budget"*. I like the way this was worded. He was demonstrating that he had direct ownership of a $2M budget, meaning he ultimately decided how the money was spent and was responsible for the value the investment provided. But what really captured my interest was the $25M reference. He understood the connection with the big picture and how he shared in the accountability of the whole amount. Big numbers are good and they will get people's attention, but make them real. You won't fool anyone if you are just plucking numbers out of thin air. I did by the way, ask him to detail the makeup of the $2M amount, how it was spent, and how he actually managed it. Again, if you have not had any financial responsibility to speak of, that's okay. Most IT organizations have centralized the P&L accountability within the PMO (Project Management Office) leaving the delivery leaders to focus on resource and task management.

Next you will describe the responsibilities of your position. This is the meat of this exercise and I suggest spending the majority of time on this piece. Later this paragraph will be used to build your resume—in step 8. Think about this for a moment and don't just simply describe the job role. Describe what you did and try to describe how well you did it. Experienced IT hiring managers already have a good understanding of what most positions in the field entail. They don't need a lesson from you on job title descriptions. Avoid statements such as the following, "Reported project budgets and milestone status to senior management." That is only stating the obvious and it wastes space. In step 8, I will discuss the idea of limited *resume-real-estate* and why you don't want to squander this valuable space with detailed descriptions of your daily activities. Nothing can be more powerful than hard facts such as, *"Contributed to the delivery of an inventory system essentially cutting a month long process down to a matter of hours resulting in operational cost savings of over $200k annually."* Now that is powerful and will get noticed. So, don't let this be a boring or generic description of your duties, make it sexy, make it powerful, *and make it sell!* Remember the purpose of this step is to focus on your achievements, rather than just describing your experience. Structure your job description statements with action-oriented verbs and phrases that will get you the attention and consideration you deserve. By the way, need help with action-oriented verbs? See the end of chapter 6 for a good list.

When answering the question, "What did you like **most** about this position?", try listing one or two of your top professional values from step 3 if they apply. Avoid listing things like the "pay was great" or "the raises were large". As honest as these answers might be, they imply that you are solely motivated by money and that you will most likely leave for a more lucrative offer later on. When answering the question, "What did you like **least** about this position?", you want to try to turn this into a positive. Don't list things like you hated having to sit in all the boring meetings, writing long status reports, or my personal favorite "my relationship with my manager was damaged beyond repair". I've gotten this one a few times, and although it may be true, it's not one to list on a job application. Focus instead on issues such as you "didn't see personal growth opportunities" or you "didn't feel professionally challenged by the job activities" or "the leadership was not focused and direction was unclear".

Refer to the talents and strengths you listed as part of step 4 when answering the question, "What were the strengths that you brought to this organization?" If your answer to this question is short, it will undoubtedly be long to the next one, "Reasons for leaving." For reason for leaving, you can try to list how your strengths were not properly applied. Try also listing a few of the lowest ranked professional values from step 3—these are the values you considered unimportant and not necessary to achieve job satisfaction.

Finally list any hard (technical) skills required by the job. You can use your answers from step 4 here. Again, ignore this one if no particular hard skills were required by the position.

Job History

Company Name: _____   **Industry:** _____

Your Title: _____ **Supervisor's Title:** _____

Company's product and/or mission.

_____

Number of direct reports: _____

Number of resources whom you were responsible for: _____

Profit/Loss (Budget amount)   **Responsible: $**_____   **Accountable: $**_____

Describe your major responsibilities: _____

_____

_____

_____

_____

_____

_____

What did you like most about this position?

_____

_____

_____

What did you like least about this position?

_____

_____

List the strengths you brought to this organization:

_____

_____

Reason for leaving: _____

_____

_____

Hard (technical) skills required by the job:

_____

_____

Chapter Six

Become a Better Story Teller

It usually takes me more than three weeks to prepare a good impromptu speech.
—Mark Twain

Storytelling evolved from a need to communicate our individual, shared or community experiences with others. Storytelling is an integral part of everyday life; it is in the books we read, the TV shows we watch, and in the web sites we surf. We tell stories to entertain our friends, to teach our children, and to share our thoughts and ideas with our peers. You could say each of us is the sum of the stories we tell about ourselves. Your ability to tell a good story will be critical during your job search campaign.

I have to admit, I learned how to tell a good story from my father. Gerry had a story for every occasion. When my brothers and I would listen to his stories we were transported, together, outside of the present moment, to another time and place. We would live the experience of the story's characters through the use of our imagination. His stories would inject emotion and passion into the content and we were tightly connected in that moment in time. He had our undivided attention for the entire length of the story.

In his best-selling book, *Squirrel Inc.*, former World Bank executive and master storyteller Stephen Denning used a tale to show *why* storytelling is a critical skill for professionals. *Squirrel, Inc.* details the art of storytelling. It's an engaging read from cover to cover filled with useful ideas that you can immediately apply to your storytelling skill-set. Effectively telling the stories of your accomplishments will be required during every interview. You must be able to deliver a tale that keeps the attention of the audience while making a strong statement of your qualifications for the job. During interviews, the hiring manager's questions will most likely be prefaced with one of the following:

> *Tell about a time when you...*

> *Give me an example of when you ...*

Describe a situation when you...

This will be your opportunity to make an impression by telling a compelling story. Your stories can be short or they can be long. A good rule of thumb is to keep your stories to no less than one minute and no more than three minutes. Three minutes may not seem like a lot of time but consider this, while **speaking at a moderate pace, you can say about 500 words in just three minutes.** Anything more than that and you are going to lose your audience's attention. It is also a good idea to have a long and a short version of each story. You'll know when its appropriate to use either version. For example, the hiring manager gets to hear the long and detailed version while his boss only needs to hear the short, high-level version.

Your stories will be real accounts of your experiences that demonstrate your skills and values and how you have handled and have overcome situations. Your exercise here is to take the first step towards weaving a good yarn. Deciding what stories need to be told and writing them down is that first step. You will prepare a collection of stories that will be part of your "toolbox" to be used throughout your job campaign. Your stories will follow a proven format that is easy to develop, is organized, structured, and gets to the point.

To structure the stories of your work experience, you will use an approach known as **SAI (Situation\Action\Impact)**. Getting started, think about experiences where you had a noticeable impact on the organization. Events that not just helped your team succeed but really achieved an organizational goal. There is a blank **SAI** form at the end of this chapter for you to use. Make extra copies so that you can record several stories with long and short versions. You can also download copies of these worksheets and others from this book's website: www.itdreamcareer.com. You will also find a couple of sample **SAI**s, approximately 500 words, that are true examples that I personally used during job interviews.

Situation

This is the area where you really set the scene. Describe the situation that you faced in your organization. Qualify the challenge, and then quantify the problem if you have the data. Describe where the organization was headed—i.e. what would have happened if you did not take the action that you did?

Action

Describe the specific actions that you took. Make sure and use action words (see list at end of this chapter). This is your chance to shine by showing what you did that made a positive impact to your organization. Don't embellish, there is no need to stretch the truth. Simply state what you did or how you led your team to accomplish your goal.

Impact

What were the results? Describe the outcome. How did you know it was a success? Did the CIO or the vice-president (or whomever may have been in authority) recognize you or your team? Did the customer recognize the result? If you saved the company dollars or man-hours or lives, definitely try to list or quantify that success. If you don't remember exact figures, then approximations are ok. Include the timing of the impact as well. Make sure you clearly describe the ***impact*** that you had to the organization—not just your department or your team. Having this enterprise mind-set will help you stand out from the crowd of applicants who are only looking at their little areas of responsibility rather than the big picture.

Review

Re-read your work to ensure it makes sense. Have you followed the stated outline of **Situation** (setting the scene), **Action** (what you did), and **Impact** (end result and benefit)? Have your spouse, friend, parent, or coach review what you wrote as well to see if it makes sense to them. Make sure you know your stories well when it comes time to tell them. Practice this. There's nothing quite as pathetic as someone who is telling a story and forgets how it goes!

You will need to build a collection of **SAI**s for your toolbox. As you do this, think about the type of job that you are trying to get and then think about the *experience areas* and *skill sets* that are most relevant to that job. From your experience, pull out successes that you have in those areas to develop your set of **SAI**s. Ideally, you will want to draft eight to ten **SAI**s that can be part of your campaign tool kit. In later chapters, we will discuss how you will use your **SAI**s, tell your stories, throughout your campaign in a way that will allow you to powerfully present yourself to a prospective employer.

SAI Sample #1

Company: XYZCorp **Position:** IT Delivery Manager

Situation:
As part of the corporate objective to replace legacy applications and implement service oriented technology that integrates well with partner systems, I was charged with managing the replacement of applications that supported the national sales division. This initiative presented a significant challenge to both the technology and business side of the house. On the technology side, I was asking application support staff to participate in the development and/or evaluation of systems that would ultimately replace the ones they have supported for decades. Including the purchase of software and support services that could eliminate the positions they occupied. On the business side, I was asking sales staff to participate in a comprehensive business process analysis and definition effort to define their functions and workflows.

Action:
Influencing and gaining commitment from potential skeptics and resisters was paramount. I had to get complete buy-in and support from the technology staff and the business community. The technology staff had old skills (legacy-mainframe), old techniques (data centric) and archaic methodologies (waterfall SDLC). The business staff had never been involved in software solution development and could not understand why they were being asked to participate. They had, through the years, grown used to the 'over-the-wall' software delivery process. I coached legacy developers on the necessity and benefits to acquiring new skills if they were to survive this enterprise transformation. I convinced sales management of the necessity and benefits of constant involvement and input from them and their staff in order to deliver solutions that would truly automate and streamline their business process.

I established formal communication forums (steering committees, focus groups, technology review teams). When businesses use the same technology for many years, some of the business knowledge gets lost in the technology (users forget why they perform certain tasks, or systems makes decision for them behind the scenes). To fully understand the business process I got the technology group and the business group to communicate and describe the processes together. I accomplished this by leveraged visual modeling tools to map the business process and produce UML artifacts (activity and use-case diagrams). These UML artifacts facilitated a common communication language between users, developers, managers and executives.

Impact:
Technology folks acquired new skills and learned new development methodologies. Construction and deployment efforts were more effective by following stricter rules and guidelines. IT resources evolved from pure programmers to programmer/analysts/integrators, they got excited about learning new technologies and techniques, which would enhance their daily activities and even their careers. Business folks reacquainted themselves with their business processes and, as a result, improved their daily routine and how they accessed and reported information. They got excited about participating in JAD sessions and Prototype sessions. Touching (seeing and feeling) applications as they were constructed over time got business folks very excited and allowed them to participate in the ongoing development (in terms of look and feel).

SAI Sample #1 (continued)

The project was a huge success due to the commitment and contribution by everyone involved. My strategy to motivate and get all stakeholders excited about the project led to a very positive outcome for the project and for the organization.

SAI Sample #2

Company: XYZCorp   **Position:** Computer Operations Manager

Situation:
As the Computer Operations Manager for XYZCorp, a $450MM company, the national support center's managers approached me about the need to capture the knowledge gained by our technicians in order to provide for faster problem resolution, prevention of repeat mistakes, and the faster training of new technicians. Consequently, I initiated actions to develop a knowledgebase.

Action:
My first action was to hold a meeting of my staff from which I solicited input. After some discussion, agreement was reached and I assigned basic research was to be done on the content and construction of a knowledgebase to one of our lead support center technicians. Once this report was completed, I reviewed the findings and I scheduled a meeting to determine and spec out system requirements for a knowledgebase. I facilitated the evaluation of two possible platforms for a knowledgebase that had been identified in the report. I needed to determine how to structure the knowledgebase in some sort of hierarchical fashion so that it would be usable to the support center technicians. I directed our Remedy specialist to develop the structure that matched closely with our existing support structure. The team then planned how to populate the knowledgebase and then implemented the project as a pilot. Once an assessment of the pilot was complete, the knowledgebase procedure was adjusted and then fully implemented. I tracked the use of the knowledgebase to determine its ease of use and value.

Impact:
Reviews of customer satisfaction during a 3 mo. period revealed that customer satisfaction increased from 80% to 95% which improved end user effectiveness and increased productivity both among the end users and the support center staff.

Action Verbs

A
Accelerated
Accomplished
Achieved
Acted
Activated
Adapted
Addressed
Adjusted
Administered
Advanced
Advertised
Advised
Advocated
Aided
Allocated
Analyzed
Answered
Applied
Appraised
Approved
Arbitrated
Arranged
Ascertained
Assembled
Assessed
Assigned
Assisted
Attained
Augmented
Authorized
Awarded

B
Balanced

Began
Boosted
Briefed
Budgeted
Built

C
Calculated
Captured
Cataloged
Centralized
Chaired
Charted
Checked
Clarified
Classified
Coached
Collaborated
Collected
Combined
Communicated
Compared
Compiled
Completed
Composed
Computed
Conceived
Conceptualized
Condensed
Conducted
Conferred
Conserved
Consolidated
Constructed
Consulted

Contacted
Continued
Contributed
Controlled
Converted
Conveyed
Convinced
Coordinated
Corresponded
Counseled
Created
Critiqued
Cultivated
Customized

D
Debugged
Decided
Defined
Delegated
Delivered
Demonstrated
Designated
Designed
Detected
Determined
Developed
Devised
Diagnosed
Directed
Discovered
Dispensed
Displayed
Dissected
Distributed

Diverted
Documented
Drafted

E
Earned
Edited
Educated
Effected
Elicited
Eliminated
Emphasized
Employed
Encouraged
Enforced
Engineered
Enhanced
Enlarged
Enlisted
Ensured
Entertained
Established
Estimated
Evaluated
Examined
Executed
Expanded
Expedited
Experimented
Explained
Explored
Expressed
Extended
Extracted

F
Fabricated
Facilitated

Fashioned
Finalized
Fixed
Focused
Forecasted
Formed
Formulated
Fostered
Found
Fulfilled
Furnished

G
Gained
Gathered
Generated
Governed
Grossed
Guided

H
Handled
Headed
Heightened
Helped
Hired
Honed
Hosted
Hypothesized

I
Identified
Illustrated
Imagined
Implemented
Improved
Improvised
Incorporated

Increased
Indexed
Influenced
Informed
Initiated
Innovated
Inspected
Inspired
Installed
Instituted
Integrated
Interacted
Interpreted
Interviewed
Introduced
Invented
Inventoried
Investigated
Involved
Issued

J
Joined
Judged

K
Kept

L
Launched
Learned
Lectured
Led
Lifted
Listened
Located
Logged

M
Maintained
Managed
Manipulated
Marketed
Maximized
Measured
Mediated
Merged
Mobilized
Modified
Monitored
Motivated

N
Navigated
Negotiated
Netted

O
Observed
Obtained
Opened
Operated
Ordered
Orchestrated
Organized
Originated
Outlined
Overcame
Overhauled
Oversaw

P
Participated
Performed
Persuaded
Photographed

Pinpointed
Piloted
Pioneered
Placed
Planned
Played
Predicted
Prepared
Prescribed
Presented
Presided
Prevented
Printed
Prioritized
Processed
Produced
Programmed
Projected
Promoted
Proofread
Proposed
Protected
Proved
Provided
Publicized
Purchased

Q
Qualified
Questioned

R
Raised
Ran
Rated
Reached
Realized
Reasoned

Received
Recommended
Reconciled
Recorded
Recruited
Reduced
Referred
Regulated
Rehabilitated
Related
Remodeled
Rendered
Reorganized
Repaired
Replaced
Reported
Represented
Researched
Reshaped
Resolved
Responded
Restored
Retrieved
Reviewed
Revised
Revitalized
Routed

S
Saved
Scheduled
Screened
Searched
Secured
Selected
Separated
Served
Shaped

Shared
Simplified
Simulated
Sketched
Sold
Solved
Sorted
Spearheaded
Specialized
Specified
Spoke
Sponsored
Staffed
Standardized
Started
Streamlined
Strengthened
Structured
Studied
Suggested
Summarized
Supervised
Supplied
Supported
Surpassed
Surveyed
Sustained
Synthesized
Systematized

T
Targeted
Taught
Terminated
Tested
Tightened
Totaled
Tracked
Traded
Trained
Transcribed
Transformed
Transmitted
Translated
Traveled
Tutored

U
Uncovered
Undertook
Unified
United
Updated
Upgraded
Used
Utilized

V
Validated
Verbalized
Verified
Vitalized
Volunteered

W
Weighed
Widened
Won
Worked
Wrote

SAI Form

Company: _____   **Position:** _____

Situation:
_____
_____
_____
_____

Action:
_____
_____
_____
_____
_____
_____
_____
_____
_____

Impact:
_____
_____
_____
_____
_____
_____

Chapter Seven

Craft Your Personal Brand

My greatest strength is common sense. I'm really a standard brand - like Campbell's tomato soup or Baker's chocolate.
—Katharine Hepburn

Fashions fade, style is eternal.
—Yves Saint Laurent

Your personal brand is what others perceive you to be. During your job search campaign, you will have several opportunities to convey your personal brand. Your resume is one way; a telephone interview conversation is another. Your presence on the web—either through social networking or simply your profile home page—can also communicate your personal brand. But no means of conveying your personal brand is as effective as your physical presentation. Presenting your physical appearance (image) in the right way will influence what others perceive to be your personal brand.

You might be a perfect candidate on paper, but a disappointment in person because of how you present yourself. When you walk into an interview where no one has ever seen you before, you create a "threshold effect." This means that in that decisive moment people will determine your educational level, your trustworthiness, your success in previous endeavors, your moral character and more. This is your moment in time to create a great first impression and first impressions are lasting impressions. You must capitalize on this effect because you will not get a second chance to make a great first impression, one that will set you apart from the competition, so make sure your *first impression is the right impression.*

There will be many opportunities for you to interact and meet other professionals during your job search campaign and that is why I felt it necessary to dedicate a whole chapter to discuss techniques used to create and showcase a wining personal brand. And you'll be happy to know there is no exercise in the chapter.

DRESS

Most of us probably follow the common rule of wearing a suit and tie or pant/skirt suit with heels to an interview. You should always dress your best, dress for success, the way you dress will make a difference in whether you get considered for the job, or not. At times you may feel unsure about what to wear and you may worry about *overdressing* especially when you know the office you're going to has a *biz-caz* dress code. Don't be unsure, dress professionally. It's always better to make a sound impression rather than risk appearing slouchy or disinterested. And as much as you Gen Y'ers like to push the limits, break the rules, and dress casual, an interview at the office is just not the place and time to make your fashion statement. If you're in doubt about how to dress for an interview, it is best to err on the side of conservatism. It is much better to be overdressed than underdressed. The better your exterior looks the more confident your interior becomes.

How you dress is your **first opportunity** to showcase your personal brand and impress the interviewer. When you walk into an interview, your image will be noticed before you even have a chance to say a word or shake a hand. I still cringe when I think of some of the outfits I've seen candidates wear during interviews. There was the technical manager candidate who showed up in a dirty old wrinkled, I mean *wrinkled*, suit, with no tie and old sneakers. The sad thing was that he was extremely talented, but for a corporate environment where he would interact with customers, I just couldn't hire him. There was the project manager candidate in a very short, very tight, and very bright pink dress, and she could barely sit down. She was dressed like she was headed out to a club rather than to apply for a corporate management position. Oh and I'll never forget the fellow I interviewed for a Java developer position who showed up in shorts. His excuse was that he came straight from the airport and didn't have time to change. Needless to say, he didn't get the job and neither did the other two. Don't get me wrong, I don't hire people based solely on how they dress. That is not what is most important to me. However, I do expect them to look decent and *show respect for the process.* Dressing too casually shows lack of concern for the company's policies and a lack of interest in the job. After all, if I can't expect you to follow a simple interview dress code, what success am I going to have expecting you to follow more important company policies?

SMILE

A warm smile is your **second opportunity** to showcase your personal brand and impress the interviewer. A warm smile is a powerful weapon—it can make others think that you have something special. Smiling will also make you feel better, give you confidence, make you more attractive, and give you that needed emotional boost during interviews. Oh and by the way, others may not notice your bad dress habits if your wear a big smile.

"A smile is an inexpensive way to change your looks." ~Charles Gordy
"Life is like a mirror, we get the best results when we smile at it." ~Author Unknown
"Everyone smiles in the same language." ~Author Unknown
"The shortest distance between two people is a smile." ~Author Unknown
"You're never fully dressed without a smile." ~Martin Charnin
"Of all the things you wear, your expression is the most important." ~Janet Lane

Okay, you get the point... SMILE!

HANDSHAKE

A firm handshake is your **third opportunity** to showcase your personal brand and impress the interviewer. This two-second ritual starts the interview and can send a message. For example; the *dead fish* (limp hand) handshake often gives the impression of weakness or disinterest in the occasion. If you only extend the tips of your fingers you might appear to be unengaged and if you rapidly pump your arm you will appear overly aggressive. In my experience as a hiring manager, all too often an interview gets off to a bad start because of an ineffective handshake. A good handshake consists of a full hand clasp, shaking up and down three or four times while using direct eye contact. So, avoid the bone crushers or limp hand—your handshake is saying more about you than you think. A firm hand shake shows your interest, eagerness, enthusiasm and energy.

EYE CONTACT

Good eye contact is your **fourth opportunity** to showcase your personal brand and impress the interviewer. Good direct eye contact will show sincere interest in the person you are meeting. It adds to your credibility and believability. A good rule of thumb is to maintain eye contact with the interviewer 75% of the time. Anything more than that and they might feel like you are trying to hypnotize them. During interviews, I've found that the more technical a resource is the more introverted the person seems to be. For shy folks, direct eye contact is particularly difficult and uncomfortable. But it can't be ignored as a powerful tool that can either help you or hurt you during the interview.

BODY LANGUAGE

Good body language is your **fifth opportunity** to showcase your personal brand and impress the interviewer. The trick is to strike a posture that demonstrates interest but still comes across as being relaxed. If you slouch or hang sideways it will give the impression of being uninterested. Sitting on the edge of your chair can come across as being a little tense and might give the impression that you feel uncomfortable. Folding your arms might feel natural for this situation, but can be interpreted as being defensive. It is better to let your hands lie loosely on your lap or place them on the armrests of the chair or try using your hands to support your words with hand gestures. Always sit up straight in your chair with your back against the back of the chair. When the interviewer speaks it's good to lean forward a little and tilt your head slightly, this shows an interest in what they are saying.

So, dress appropriately, extend a firm handshake, maintain good eye contact, be careful of your non-verbal body language, and most importantly *smile*. Follow this simple five point formula to make a good and lasting first impression because your personal brand depends on it.

Chapter Eight

Build an Effective Resume

Do you know the difference between education and experience? Education is when you read the fine print; experience is what you get when you don't.
—Pete Seeger

If you do a Google search on "resume", you will get more than 75 million search results. The results include links to web pages offering professional resume services, websites selling books on crafting resumes, and hundreds of knowledgebase sites offering resume writing tips and strategies. Narrow down your search criteria to just "IT Resume", and you will receive over 50 million search results. These sites provide best practices on how to create a good IT resume. Finally, narrow down your search criteria to "make the perfect IT resume" and you'll still receive over 10 million search results. The point is, there is an amazing amount of information available, most of it free, providing you the basic guidelines to crafting a good resume. But following the basics won't separate you from the pack, and therefore won't be enough if it is to get you to the next step—an interview.

You need an edge...

I mentioned in the last chapter that your resume is another opportunity to convey your professional brand. And in most cases, your resume will be the first thing that your potential employer reads about you. That is why it must be more than a dry chronological history of your career; it must communicate your past success and your future potential. It must peak the interest of the reviewer and make them want to talk to you and find out more about you. Think of your resume as your personal marketing brochure designed to persuade prospective employers that you will be a perfect fit for the open position. And to do this you must extract the relevant accomplishments from your experiences and highlight them compelling the hiring manager to not only want your services, but to *need your services*.

I'm guessing you are reading this book because you are out of work or unhappy with your current position. Reflect for a moment to a time when you were happy at work. A time when you had a "great job". You didn't look for a job then, because you were happy. But I bet if you had looked for one, you would have been hired instantly. Why? Because people want to be around positive people. You've heard the expression, "it's easy to get a job when you have a job", how ironic and true. When you have a good job, you are confident and **passionate**, your attitude is positive and optimistic, you feel like a winner and it shows.

The exercise in this step is to craft a winning IT resume. One that has energy, personality, and one that stands out from the crowd. Oh by the way, all that work you did in previous steps, like identifying your talents, documenting your career history, and telling your stories can now be put to good use in helping you create a knock-'em-dead resume.

Quality Assurance

I cannot overstate the importance of proof reading your online job applications, job search emails, and above all, your resume. You must always check the spelling and the grammar on your resume. I'll be the first to admit, I'm a bad speller—seems to be a common trait among us *techies*. There are times when I'll be reading through a good resume thinking to myself I have found the perfect candidate for this position. Next thing I know, up pops a blatant misspelled word or badly structured sentence. So I let it go, after all nobody's perfect and I'm not looking for a book editor. So I keep reading. A few seconds later, another misspelling and then another bad use of grammar. Now my focus has shifted away from the candidate's qualifications and over to the mistakes, as I am now waiting for the next typo to pop up. You might think I'm a bit picky. I'm not. Your resume must be 100% free of mistakes. Hiring managers, while reading your resume, are not going to allow for one single misspelled word. Your resume is only a few pages long; if you can't even take the time to spell-check and proofread such an important document, then why should you be considered for the job? It's simply careless and unprofessional. Your resume is a reflection of you and your attention to detail and your ability to create a quality work-product. Ninety-nine percent of professional positions depend at least somewhat on your ability to write. Your resume is the first chance (and possibly only chance) that the hiring manager will get to evaluate you on this skill before making a hiring decision.

Every word processor has spell-check. Use it. But don't stop there. Spell-check will only flag words that are misspelled and sentences with basic grammar mistakes. Spell-check won't catch the misuse of technology acronyms and abbreviations. For instance Websphere vs. WebSphere, Web services vs. Webservice, or RBDMS vs. RDBMS. For your IT resume, you will have to do spell-check the old fashioned way. Look it over, read it out loud several times. Remember when I mentioned, in chapter one, the value of having a campaign partner? Well this is where they come in. Give them a copy of your completed resume and ask them to read it and look for mistakes. Guaranteed they will spot something you have missed. It is awfully difficult to objectively proofread your own document. One resume that I did for myself was "proofread" by me over a dozen times. When I asked my wife to review it, she still found an obvious typo that I had completely glossed over a dozen times. After that, I learned never to trust only myself in proofreading my own document. A mistake free resume will ensure the hiring manager's attention is focused on your great work achievements and position qualifications and not on those pesky typos.

Don't Get Cute

In case you're not sure, it is not appropriate to include a photo of yourself on your resume. I know we are all WORD experts and inserting an image on the upper left or right-hand corner is easy. Resist the urge. I remember getting a resume for a senior level architect position with a photo on the first page. What makes this image stick in my mind is the fact that it was an image of him sitting in front of a computer cranking away, grinning from ear to ear wearing an '80's hair band concert shirt. No pictures please! Just stick to the basics. Use a standard black color font on plain white paper. No icons, images, backgrounds, or fancy bullets.

Don't Lie

Most recruiting firms and employers do extensive background checks on candidates prior to an offer of employment. Basically, the process verifies the information on your resume and/or your application. Most often these verifications happen during the interview process (requiring your authorization). Some employers have hired companies that specialize in background checks and they are very good at finding the truth. It used to be that if you wanted to get ahead, you didn't think twice about padding your resume with certain untruths. Maybe you included a degree you didn't exactly earn or you claimed to be the captain of the university lacrosse team or you listed a job title you didn't really have. Avoid the temptation to lie or embellish experiences on your resume. If you do lie, be prepared to get caught. And even if your lies get past the screening process, you will eventually be asked to account for the job experience. As a hiring manager, I always made it a point to flush out the details. If it became obvious that the information on the resume was an exaggeration, I immediately shortened the interview and passed on the candidate. Even if I felt he or she was qualified, I passed. The way I see it, if you don't think twice about lying on your resume, chances are you'll cheat in your daily work activities.

Don't Tell Your Life Story

Your resume should not be more than two pages long. Let me say that again, *your resume should not be more than two pages long*. I see a problem when a candidate cannot adequately communicate the necessary information in two pages. I immediately think the candidate has had too many jobs, or has a career that is not focused or is simply unable to be concise when appropriate. The truth is, as a hiring manager, "I hate long resumes". They encourage me to pass on the candidate and move quickly to the next resume. My "long-resume" record as a hiring manager was a 13-page word document. I must admit, it got a few laughs in the office, but then went right into my trash. And consider this if you *must* create a long resume, a resume that *lists everything you have ever done* requires that you are prepared to *discuss everything that you have ever done* when you are interviewed. Do you really want to put yourself in that position?

In chapter five, I mentioned *resume-real-estate* and that you must not squander this valuable space with detailed descriptions of your daily activities. No matter how tempting it is to go into detail about every job you had over your 20 year career, don't. What you did in IT over 10 years ago is irrelevant. Technology platforms have changed, approach to technology has changed and how technology is delivered has changed. If you are a seasoned technology resource, then a brief account of your history will suffice. Let the lion's share of your two-page resume showcase your most recent and relevant accomplishments. Think of it this way, a hiring decision maker is not gauging whether you are a viable candidate for that CIO position based on your first Cobol programming position out of school 20 years ago.

I typically review dozens of resumes for any one particular position. I spend between 30 and 90 seconds on each one. And that's it. You either make it or you don't in less than 90 seconds. My goal in this short period of time is to get a sense of the candidate's background and what they have accomplished with past organizations. At the same time, I am trying to determine if their experience meets the qualifications of the open position. The only decision I will make after reviewing each resume is whether or not to put the resume on the *next step* stack or into the trash. Keep your resume to two pages and focused on recent achievements.

Resume Format

There is no one standard format for a resume and as such, successful resumes (i.e. those that have been used by people who got the job) vary widely in format.

The three most common formats include:

- The Chronological Resume
- The Functional Resume
- The Accomplishment Resume

The **Chronological Resume** is probably the most common format, especially for people who have a substantial work history. I am recommending you use this format. This style emphasizes your work experience and past history. The format is "chronological", meaning that your professional experience is detailed in reverse order (most recent position first). Remember, your most recent position is the most relevant.

The **Functional Resume** emphasizes your skills and accomplishments. It presents a summary of your past work experience, linking it to the skills and achievements you present. This style might appeal to you if you have held jobs in a lot of similar positions. Rather than having to list each individual job, you can save room by focusing on an overview of your work history. This style removes the redundancy.

The **Accomplishment Resume** does not list your previous positions and experience by date, but rather lists accomplishments in no particular order with no date or organization attached. You might use this resume format if you have been out of the workforce for a number of years or are changing careers. I recommend you avoid this style, as it may raise more questions than it answers because it looks like you might be hiding something.

Again, I'm recommending you use the Chronological resume format. It lends itself very well to our industry (IT), it is simple to create and easy to read.

Build Your Resume

Before we begin this exercise, go back to chapter six and review your stories that describe all the positive things you achieved at your previous jobs. These stories will be the essence of your resume. We'll use the sample resume later in this chapter as our guide.

There are 5 sections in your resume:

1. Header with contact Information
2. Summary of qualifications
3. Professional experience
4. Education and certifications
5. Additional qualifications (optional)

Section 1: Header with Contact Information

Your name and vitals go front and center right at the top. The font for this section should be slightly larger than the rest of your resume. For this section I recommend font size 13-14. You want this information to stand out so the hiring manager can easily find your contact information when scheduling your interview. The remainder of the resume can be font size 11-12.

The following example is a simple and effective header:

Dewey Snow
9005 Sandpiper Court – Orlando, FL 32888
Home (000) 555-5555   Mobile (000) 555-4444   DeweySnow@gmail.com

Section 2: Summary of Qualifications

The structure of this section is meant to be a powerful, hard hitting representation of your background and experiences—presented in a way that will compel the reviewer to want to know more about you. You will want to write a good summary paragraph of your background, experiences, and qualifications. To help pick the right words, do a Google search on "resume action words", or review the list at the end of chapter 6. Here are a handful of summary paragraph key words and phrases which I found to be very effective:

- Team oriented, proactive executive
- Disciplined, organized professional
- Organized team builder
- Decisive, confident professional
- Assertive, confident professional
- Disciplined, energetic professional
- Energetic, enthusiastic professional
- High-energy, aggressive professional
- Personable, energetic professional
- Seasoned executive
- Creative, disciplined professional
- Innovative, tenacious professional
- Confident and persuasive professional
- Innovative problem solver
- Persistent, innovative professional
- Results driven executive
- Competitive, disciplined professional
- Confident self-starter
- Organized, effective professional
- Dedicated, innovative professional

Here are 2 examples of an effective summary paragraph:

Example #1:

Confident self-starter with twenty plus years of global experience in information systems and project management with increasing management responsibilities. Organized team builder with experience managing global cross-functional teams. Excellent program and project management skills (e.g. tools, methods, expectation setting, relationship management, multi-project programs). Industry experience includes Automotive Manufacturing, Energy, Financial Services, Hospitality, Supply Chain Management, and Telecommunications.

Example #2:

Creative, disciplined Senior Software Engineer / Architect specializing in J2EE, Java Web Application, XML and Web Services (SOA) application development. Innovative problem solver with project leadership and development lead experience. Complete multi-tiered application development lifecycle experience using J2EE standards. Sun Certified Programmer for the Java 2 Platform 1.5. Over 10 years of object-oriented analysis, design, and development experience. Solid UNIX, Linux and Open Source development background. Current with industry best practices and technologies by following weblogs, forums, mailing lists and open source projects.

Section 3: Professional Experience

This section gets to the meat of your resume. It lists each employer and position in reverse chronological order (most recent first). In a tight, bulleted format, you want to list what you did and how it positively impacted the organization. Quantify if possible, but descriptive words are fine to use as well. Refer to your **SAI** stories completed in chapter 6. You should pull from those stories and briefly describe them in this section of your resume. Your campaign will have a consistent theme, based on your experiences and the value that you brought to each employer and in each position that you have held. Your collection of **SAI** stories represent the thread of your theme that weaves through your job search campaign. You will be writing and talking about your stories in your resume, in your interviews, in your elevator speech, and during your entire job search.

Here is an example of an effective professional experience section:

Senior Technical Support Analyst
Responsible for the strategic development and implementation of cost-effective training and support solutions that were designed to provide improved productivity, streamlined operations, and faster access to critical information.

- Implemented effective customer satisfaction strategies by identifying and eliminating the root causes of customer problems.
- Utilized custom application to manage call center metrics, led call calibrations, and performed random-sample audits on email and chat sessions.
- Handled escalation processes and mentored other support professionals while working via phone, email, and chat.
- Provided comprehensive system support, configuration, maintenance, and training for Meridian Bank and promoted value added products and services for existing clients.

You need one of these for each of the last 2 - 4 jobs. Use the next page to write out your professional experience paragraphs. This is just a first draft that you will later have a chance to refine and smooth out. For now, all you need to do is brainstorm, reflect back on your experiences, use your stories and write.

Section 4: Education and Certifications

List your degree and major along with your school. No need to list graduation dates or GPA or any other information that is not directly relevant. If you have a Minor that is relevant to your job search, then go ahead and list that as well. Make sure and spell out the degrees—don't abbreviate. This section may also be used to list certain technical certifications which are common in information technology.

Here is an example of a simple and effective education section:

Bachelor of Science, Computer Science, University of Central Florida.
Master of Science, Information Management, Rollins College.

Certified Project Management Professional (PMP) Certification

Certified NetWare Engineer
Microsoft Certified Systems Engineer
Certified Six Sigma Black Belt

Section 5: Additional Qualifications (Optional)

List any additional certifications or affiliation you think are relevant to your job search.

- Member of the Project Management Institute
- Women in Project Management Special Interest Group
- Senator Representative for Political Science Department
- Department of Defense Security Clearance - Granted: Nov 2005
- Orlando Java Users Group
- Orlando .NET Users Group
- President of the Local Chapter of Toastmasters

References

One thing you notice is NOT shown on your resume, is a list of references. At some point in your campaign you will indeed be asked to provide references—typically former supervisors, bosses, or colleagues—who can vouch for your work habits and experiences. However, there is no need to provide your references up front in your resume. Instead, you want to have a set of references available for a prospective employer who asks for them. Asking for references is a good sign because it shows that the prospective employer is taking a good look at you.

When building your reference list, think about people you have worked for or worked with in the past and who you feel would give you a good reference. Call them up (or send them an email) and ask their permission to use them as a reference. By the way, this is a great way to perhaps get back in touch with someone if you have not spoken with them in a while. They might be interested in the fact that you are "in the market" for a job and may have something available. If you are currently employed but are "looking", you can ask the reference to keep your search confidential. A typical person will be more than happy to serve as a reference. Then, each time you give their name out to a prospective employer, make sure you let your reference know that as a common courtesy. Also, try not to overuse your reference, i.e. don't give their name and contact information to dozens of potential employers. Probably best to use them no more than three or four times. If you have a reference letter, that is even better because you can use that several times and that will typically be sufficient for most employers although some may have their own questions and will want to speak with the reference over the phone.

Final thoughts

The resume that you have prepared is your "baseline" resume. What this means is that you will have to adjust it and tailor it for each position that you are seeking. In some cases, if the position description is somewhat vague or generic or if you are posting your resume online without regard to a specific posted position, using your baseline resume is perfectly fine. However, for sending your resume to a prospective employer in response to a specific position, you will want to make some tweaks and wording adjustments to ensure that you are communicating your qualifications for the position that they are advertising. For example, if they are looking for someone with a particular skill set and you have that skill set yet you have not discussed it in your baseline, go back and take something out that is of lesser importance and write a new version of your resume that covers the experience they are looking for. The new version will probably be 96% similar to the baseline, but tailoring your resume in this way to the specific position will make it stand out as having more impact. Also, if someone is screening your resume to ensure you have the right qualifications, your tailored resume will pass muster whereas your baseline may not.

I will have more to say about this in chapter 9 (Internet Search Strategies), but I wanted to note here that this resume style is formatted in a functional yet aesthetic way. In other words, this resume format is effective in communicating your experience while at the same time it looks good and is easy to read because it employs white space and some break-out formatting techniques. However, you will undoubtedly need to "upload" your resume to online search engines such as Monster and also to company websites that employ programs like Taleo. The format of this resume is not so conducive to the "text only" formatting required by some online job boards and job application sites. So my suggestion is that you should also prepare your resume in a text only format that still looks decent and effectively captures your information.

If you don't feel positive about one or more of your past jobs you'll have to pretend you do. Try to think of your past positions (especially the ones referenced in your resumes—the ones you'll be discussing) as fantastic opportunities and great experiences. As I mentioned earlier in the chapter, the single most important trait in an employee is attitude, and a bad one will cover your resume like a foul stench. The hiring manager won't be able to throw it away fast enough. A positive attitude toward past jobs will help you feel better about what you've accomplished and who you are, and it will show throughout your resume.

DEWEY SNOW

9005 SANDPIPER COURT – ORLANDO, FL 32888
Home (000) 555-5555   Mobile (000) 555-4444   DeweySnow@gmail.com

---

A results-oriented leader with 10+ years of IT solution delivery including 5+ years IT management experience in the following industries: Finance, Insurance, Hospitality, Healthcare and Media. Proven success at leveraging technology for competitive advantage and to support corporate goals. Knowledgeable and experienced with all phases of the software development life cycle. Demonstrated ability to drive high quality, technically complex initiatives from conception to delivery. Exceptional leadership and partnering abilities within IT and the business. Strong ability to articulate technical concepts - issues to executive management while presenting solutions in business terms.

PROFESSIONAL EXPERIENCE

SUNSHINE WORLDWIDE RESORTS - MIAMI, FL   2005 – 2008
Director, Information Technology

Developed, implemented, and maintained an information technology vision that was aligned with Sunshine's company business strategy. Responsible for a large team of 25+ individuals (both business and technical) focused in 3 areas: property systems application development, online owner self-service functionality (websites), and financial applications incremental enhancements and break fix support. Heavy focus on online reservations, membership servicing, and inventory control systems, all geared towards effectively servicing the company's membership.

- Responsible for a $5M+ annual capital budget and accountable for a $50M approved technology capital budget.
- Implemented a 'gated funding' approach to improve financial governance around project estimating and spending.
- Evangelized the use of external solution providers to meet demand and extend IT's technical capabilities and capacities.
- Collaborated with the PMO to build and enforce good PM discipline by refining the functional requirements definition and documentation process.
- Delivered a multimillion dollar inventory control system, essentially cutting a month long process down to a matter of hours.
- Delivered a specialty business servicing application providing online owner priority reservation processing.
- Supported division growth by providing technical leadership for projects related to Sales Center and Resort openings.

- Contributed to the advancement of IT governance practices by delivering functional and compliant products to the business.
- Successfully partnered with compliance to enhance audit capabilities reducing the risk of exposing PCI & PII data.
- Maintained a high performance, self-motivated work group accountable for established goals and objectives.
- Develop an effective succession plan by identifying high performers. Mentored, coached and developed these future leaders.
- Participated and contributed to technology steering committees, specifically Resort support, Web and Sales and Marketing.

APPLE TREE BROADCAST COMPANY - NEW YORK, NY   2002 - 2005
Senior Manager, Information Resources

For ATBC National Television Sales, developed the technology vision and strategy for the construction of enterprise applications to support a rapidly growing business with an increasing number of non-traditional process functions. Managed a $5M+, 12 month initiative that included several 3rd party software products integrated with internally developed applications. This solution enabled ATBC TV Spot Sales to seamlessly interface with partner applications representing 125 Advertisement Agencies and 50+ media properties.

- Operated as a Change Agent, established formal communication forums (steering committees, focus groups, etc.).
- Influenced and gained commitment to project plans and ideas from potential skeptics and resisters.
- Selected 3rd party software/service providers. Managed vendor performance, contract compliance, and SLAs.
- Leveraged Rational Rose visual modeling to map the business process and produce UML artifacts (Use Case).
- Collaborated with architects to ensure all software components were delivered in accordance with enterprise standards.
- Partnered with business unit Vice Presidents to understand and represent their unique business process.
- Reported project budgets and milestone status to the CIO of TV stations and to the President of National Sales.

GREAT BENEFIT HEALTH INSURANCE COMPANY - NEW YORK, NY   2000 - 2002
MANAGER, TECHNOLOGY DELIVERY

Responsible for the complete implementation of several eBusiness initiatives to deliver reusable, web-enabled software components on multiple platforms (.Net (C#), J2EE (EJB)). Application portfolio supported all four corporate verticals: employees, providers, brokers, and subscribers. Projects included complete Portal construction, B2B and B2C functionality, enterprise Single Sign-on (SSO), identity management, and Portal authentication and administration.

- Managed a $2M annual budget and reported costs, progress and status to senior management & steering committees.
- Directed 6 project teams (25+ resources) including Project Managers, Business Analysts, and Technical Resources.
- Identified outsourcing needs (onshore/offshore). Negotiated vendor contracts. Managed vendor relationships and SLAs.
- Managed Java developers, security architects, WebSphere administrators, and vendor technicians.
- Delivered dialog processing between Portals and back office applications (Lotus Notes, Oracle/PeopleSoft, Siebel)
- Migrated several legacy systems to web-based systems using WebSphere J2EE 3-tier architecture.
- Delivered an enterprise self-service registration, authentication and administration solution serving 1 million+ users
- Designed materials using Unified Modeling Language (UML) in the context of the Rational Unified Process (RUP).
- Produced a white paper (roadmap) to move enterprise development from WebSphere J2EE to Microsoft .NET.
- Delivered a .NET infrastructure and application consistent with the CTO's vision of migrating from a J2EE environment.

INDEPENDENT CONSULTANT - NY & NJ,   1997 – 2000
PROJECT LEAD/SENIOR DEVELOPER

- Secure Financial Group; Constructed and delivered an Asset & Liability Management System (C++ UNIX to VB/SQL.).
- Moore Medical:, Developed and delivered a Prescription Card Tracking System to 50+ desktops (100K cards daily).
- Anderson & Anderson, Delivered a Sales and Logistics tool used to evaluate and forecast drug store sales (VB SQL).
- NCI Pharmaceutica: Analyzed the business impact and risks associated with the U.S. deployment of clinical trial applications developed in Basel, Switzerland. Developed a Visual Basic application used in global budget forecasting.

EDUCATION

Master of Science, Information Management - Stevens Institute of Technology.
(Thesis: Successful Outsourcing Strategies)
Bachelor of Science, Computer Science - Rutgers University.
Certificate in Enterprise Application Development - The Technology Institute.

The next resume sample (starting on the following page) is also a good example of a concise resume that is engineered to convey specific experience with impact—prompting the hiring manager to want to know more about you.

These sample resumes may appear to be more than two pages because of the smaller trim size of this book, however, they actually fit on two pages of standard 8 1/2 x 11 paper, the recommended maximum length for a resume.

John Q. Public
1212 Main Street
Anytown, TX 55555
123-555-1212 (home); 123-555-1221 (mobile)
Email address

Innovative, disciplined, hard-working professional with more than 15 years of hands-on technical experience. Exceptional background in managing network operations, installation, and upgrades. Proven ability to analyze technology needs, develop strategy and implement technology solutions. Skilled at policy and procedure development, standardization, and project management. Consistent history of producing the right results on tight schedules.

- Team building and leadership
- Technology project management
- Systems and network management
- Network architecture and design
- Software development (J2EE)
- Systems analysis and documentation
- Technology strategy development
- Disaster recovery planning

EXPERIENCE

PRODUCT MANAGER – XYZCorp 2008 - 2010
Reported to the Vice President of Operations and assigned responsibility and oversight of all systems solutions and development efforts for the XYZwidget product. Serves as evangelist within the IT Group, the business unit, and other stakeholder groups for technology solutions to support the company's strategic goals in making the product useful and successful for XYZCorp customers.

♦ Software Development. Led a team of business analysts and programmers in the complete lifecycle development from inception phase to post-production phase of a web based reservations system to service the product customers. Successfully drafted a $2MM budget, hired staff and oversaw all aspects of the project. Project was deemed a company strategic objective with visibility and interaction at the highest corporate levels.

♦ Drives product development priorities. Developed technology strategy and project managed successful implementation of the telecommunications and IT solutions for three sales centers responsible for $70MM in annual revenue.

♦ Experienced vendor manager. Responsible for the research, evaluation, and recommendation of a customer relationship management system for the sales and marketing staff. After selecting vendor, successfully project managed the solution implementation and ongoing vendor relationship.

MANAGER, APPLICATIONS INTEGRATION – XYZCorp 2006 - 2008
Responsible for the implementation and integration of technology solutions across the corporate enterprise.

♦ Implemented controls and processes that ensured the requirements of the Sarbanes-Oxley Act were followed within the Information Technology Group. Worked with a team of IT managers to develop

policies and processes for change control and separation of environments.
♦ Planned and implemented a software development project that checked and enforced the do-not-call requirements at the corporate telemarketing centers.
♦ Evaluated vendor solutions to meet Patriot Act compliance mandates. Implemented technology based software application solution to ensure Office of Foreign Assets Control requirements were met and business risk was minimized.

MANAGER, TECHNOLOGY PRODUCTION – XYZCorp   2004 - 2006
Orlando, Florida
Reported to the Director of Information Technology. Responsible for the planning and implementation of information system technology solutions to support the sales and marketing operations.
♦ Implemented a system of metrics that evaluated software development production and assisted in the planning of software development projects.
♦ Planned and supported a successful major sales and marketing system upgrade involving 18 internal technology staff and over 200 end-users. Served as business communications liaison to ensure project objectives were planned, evaluated and effectively communicated to all stakeholder groups.
♦ Planned and implemented the technology infrastructure to include servers, computers and network infrastructure at the division's new sales center.
♦ Managed software development staff on projects to code legal documents within the division's sales system.

PROJECT MANAGER, IT – NewCorp   2003 - 2004
Orlando, Florida
Reported to NewCorp's Vice President for the Western Region and assigned responsibility for special IT projects:
♦ Evaluated IT architecture and technology solutions for Western Region resort sales and marketing and financial lending operations. Successfully deployed application servers for new project implementations.
♦ Led a team which crossed all functional boundaries in IT and operations in the planning, coordination, and deployment of a property management system significantly increasing reliability and efficiency in operations.
♦ Led a team in the planning, coordination, and deployment of an enterprise SQL Server 2000) based sales and marketing system.

COMPUTER OPERATIONS MANAGER – NewCorp   2001 - 2003
Orlando, Florida
Responsible for the day-to-day computer network operations of NewCorp's nationwide information systems infrastructure which included over 50 NetWare file servers and over 50 Windows application servers. Managed a staff of 25 technicians including the Help Desk, purchasing, and field support.
♦ Coordinated the capacity planning for our enterprise Oracle database based on HP 9000 servers using HP UX.
♦ Developed a system of metrics to measure the efficiency and effectiveness of the Help Desk staff

resulting in $140,000 in labor saving cost reduction.
- Re-engineered the purchasing process to gain efficiency and estimated cost savings of $200,000 over a six month period.
- Project managed the rollout of NewCorp's Oracle based enterprise sales and property management system to 5 resorts.
- Deployed and maintained frame relay based WAN using Cisco routers.

MANAGER, TECHNICAL FIELD SERVICES – NewCorp
Carlsbad, California
Responsible for NewCorp's network and computer operations in the western region which included 25 NetWare file servers, 16 application servers and 800 workstations. Managed a team of 12 technicians:
- Established WAN (frame relay) and LAN (Switched Fast Ethernet) infrastructure for the NewCorp office in Las Vegas, Nevada.
- Designed and built the network infrastructure and established connectivity for the Roundhill, Nevada sales and administration office.
- Designed, planned, and directed the buildout of the information systems infrastructure for NewCorp's Steamboat, CO resort.

EDUCATION

***Master of Science in Information Systems**. Graduated with Distinction from Hawaii Pacific University.
***Bachelor of Science in Physics**. Graduated with Merit from the University of Illinois.

Chapter Nine

Optimize Your Digital Search

There's a statistical theory that if you gave a million monkeys typewriters and set them to work, they'd eventually come up with the complete works of Shakespeare. Thanks to the Internet, we now know this isn't true.
—Ian Hart

The Internet has changed the way we communicate and receive information. Today the Internet is as widely used as the telephone and the television and may soon replace both of them. Just about anything you need can be found be in cyberspace, including a seemingly endless supply of job opportunities. The Internet is a gold mine of employment resources and all of this information is right at your fingertips including thousands of job related web sites all listing thousands of job openings. A successful job search campaign requires that you explore and exploit these resources.

Luckily for you, the majority of internet job sites target IT resources and most of them are free of charge. Some do charge a fee, however, I recommend you don't pay and stick to the free ones—not because the fee-based sites won't help, but because the free ones cover the majority of the opportunities. There are basically four types of internet job websites: **Corporate Websites**, **Job Boards**, **Pay For Performance** and **Social Networking.** The objective of this chapter is to provide you with the basic guidance on how to leverage the Internet for your job search. The exercises in this chapter will help you organize and keep track of your activities.

Corporate Web Sites

It can be very effective to target corporations that you want to work for. Start by finding out what companies are located in your target area and then go to their websites. Most companies have a career section where they post new positions and allow you to apply online. This is a very inexpensive recruiting method for the company which is why they will often post a new position on their own web site for a week or so to see if they have any suitable candidates before they post to one of the major job search engines.

The first step in this process is to find out what local companies are in the area you are targeting. If you already live in your target area, that should be pretty easy since you probably already know about many local companies, although there may be some that you are unaware of. When you drive through the business districts in your area, take note of the companies that you see, and then go to their web sites to learn more about them and any potential job opportunities. Also, read the business section of your local newspaper to find out stories of interest on local companies. Often you will read about who is growing and hiring, or who might be doing the opposite. There is a lot of intelligence you can gather this way that will help you target your campaign.

Another way to find out the names and web site addresses of local companies is to go to the local Chamber of Commerce web site and click on their membership directory. Don't know the website of the local Chamber of Commerce? Just Google: "Chamber of Commerce" and the name of the city you are looking at.

To get you started, focus on five to ten companies in your target geographic area. List their names and web site addresses:

Company Name   **Website**

1. _____   _____

2. _____   _____

3. _____   _____

4. _____   _____

5. _____   _____

6. _____   _____

Go to the career sections and add that to your "favorites" in your web browser. Set up a routine where you check each company's career section every day to see what new postings they may have. Some of the sites are setup with alerts, which will allow you to register and will then notify you when a matching position becomes available. Companies love qualified candidates who come to them this way because they don't have to pay expensive recruitment fees. Make sure to read the job posting carefully so you can tailor your resume and highlight your qualifications that show you are a good match for the position. Sending in your resume the same day the posting goes up on their website will significantly increase the chances that your resume will be reviewed by a decision maker. In my experience as a hiring manager, the first two dozen resumes got my full attention. That is, I read them and moved the qualified candidates to the next-step pile—typically an HR phone screen. Once I had a dozen or so candidates, I stopped reading the incoming resumes for while. The point is, you want to be one of the first responders when the positions are posted.

Job Board Sites

A job board or job search engine is a website that facilitates job hunting for individuals and candidate hunting for employers. Once registered you will be able to deposit your resume and apply for openings. Employers will have different access enabling them to post job ads and search through resumes for matching candidates. According to comscore.com (comScore Media Metrix) the most visited job boards during early 2009 were **CareerBuilder, Yahoo! HotJobs**, and **Monster.** These three sites are typically referred to as the Big Three. All three allow you to register and set up a job search agent. With a search agent, you will get an e-mail whenever a new job is posted that meets your specific criteria. To avoid becoming inundated with useless job postings which do not meet your skills and interest, make sure your job search agent is specific. Also be sure to tweak and refine the search agent along the way.

CareerBuilder (.com)
There is a reason why CareerBuilder is one of the top job board sites. In addition to having some of the best Super Bowl commercials, CareerBuilder is one of the most intuitive and well-managed job search sites out there. Through a simple-to-use interface, you can create a profile, post your resume, and begin your job search in minutes. The job search functions are extremely easy to use for both new and seasoned online job searchers. You can set automated online job search match alerts based on search specifications. These alerts can be sent daily, weekly, or monthly to your email. CareerBuilder even offers a live chat to assist you with any job search concerns you may be facing. I find the site very easy to use. The site provides many helpful career tools such as a Salary Calculator. And best of all, it's free! I highly recommend making CareerBuilder your *number one* internet job search site.

Yahoo! HotJobs (yahoo.hotjobs.com)
The HotJobs website has been around for years and is now powered by internet giant Yahoo.com. The partnership with Yahoo has increased the capabilities and effectiveness of what was already a well-established and reputable online company.

And like CareerBuilder, it's easy to use. All you have to do is create a profile, submit your resume, and begin an easy and intuitive job search experience. And also like CareerBuilder, everything on HotJobs.com is completely free of charge. You will have access to thousands of articles that offer tips on a variety of career-related aspects. You can even compare salaries not only of different careers, but of salaries in different geographic areas, as well—so you'll know if relocating to a nearby city really will provide you with more money. I highly recommend making HotJobs your **number two** job search site. Because even though it provides a lot of the same functionality and exposure that CareerBuilder does, some employers may choose to advertise exclusively on HotJobs, so you might be able to find an employment opportunity here that you won't find there.

Monster (.com)

Monster.com is very similar to the other two job boards sites and is also free. Monster has some specialized services such as Monster Mobile. The ability to search latest jobs in real time right on your cell phone—I talk about mobile apps later in this chapter. Like the others, Monster provides the ability to setup specific job search preferences associated with an industry or job role with alerts being sent on a set frequency to your email. I'm ranking Monster number three because I find the online job search to be less intuitive that the others. And even though there may be overlap, in terms of opportunity, with all three sites there will be job opportunities posted on one and not on the others. So it's definitely worth your time to create a profile and post your resume on all three. With that, I recommend you make Monster your **number three** internet job search site.

Dice (.com)

Dice.com is the premier site for Information Technology professionals. Dice has a straight forward interface and is IT industry specific. On Dice.com you can get advice and tips on getting technical certifications and there is even a discussion forum where you can talk with other IT professionals and job seekers. Dice does have a few slick features the others don't have. For example your can subscribe to Really Simple Syndication (RSS). RSS is a way for content publishers to make news, blogs, and other content available to subscribers. You can view RSS content in your email client (such as Microsoft Outlook). Using Dice RSS, you can subscribe and have information about the industry and job opportunities pushed directly to your feed reader (email) on a frequent basis.

I'm ranking Dice as one of the top five job boards sites, but not making it number one even though it is dedicated to only technology jobs. Here's why: many employers have a centralized recruiting team that responds to resource requests for many departments within the organization. The recruiters may only have an account with HotJobs or Monster and therefore will only use those sites for all job postings including the IT jobs. You definitely want to create a profile on Dice.com but make sure you are registered on the others as well. I recommend you make Dice your **number four** internet job search site.

SimplyHired (.com)
I'm adding a site called SimplyHired.com to the list. SimplyHired is a little different than the typical job search engine. SimplyHired is what is known as a vertical search or "metasearch" engine. This type of search engine allows jobseekers to search across multiple employment websites. The intention is to provide a "one-stop shop" for jobseekers who would rather have a single engine working for them rather than dealing with multiple engines. The problem is, the other job boards are wise to this strategy and are now blocking the process. Monster.com specifically bans scrapers through its adoption of a robots exclusion standard on all its pages. Still others have embraced them. This is why I'm recommending SimplyHired as one of your top 5 job search sites (because of its unique ability) but only to complement the others. This way you will have all your bases covered. Make SimplyHired your **number five** internet job search site.

Focus on these five Job Board sites. Register for each and create a profile. Create several job search alerts. Then complete the exercise by jotting down your profile credentials (username and password) and the version of your resume that you uploaded to each site.

Job Site Name	My Credentials	Resume Version
careerbuilder.com		
yahoo.hotjobs.com		
monster.com		
dice.com		
simplyhired.com		

The real power of these Job Board sites is that you can post several resumes without any fees and you can make those resumes searchable for any company that wishes to plug in various key words and search criteria. Since company recruiters will search on *key words* that are used in your resume, make sure to mention any relevant expertise and certifications (i.e. ITIL, Six Sigma, SEO, JSON/REST, CMS, etc). This should already be done in your resume, but it would not hurt to run a word search yourself to make sure you are covered. Also, look at the job postings that are already out there. What "buzz words" do you see recurring in those postings? Try to capture some of those, and include in your resume.

With each site, spend some time creating your search alerts. It is critical that you are aware of new postings in a timely manner so that you can act quickly. It is not unusual for a new job posting to generate hundreds of responses within the first couple of days. This causes the hiring company to suspend acceptance of any additional applicants until they can wade through the first batch of responses. As I said before, solid candidates are usually found in the first wave of resumes. If you don't act quickly to get in the first wave, you may not even get considered.

Pay For Performance (PFP)

There are several job search websites that are not free. They charge a membership fee and provide jobseeker services. Some of these sites are more focused on the higher paying position and target executive level job seekers. One example, and one of the more popular, is TheLadders.com. The Ladders offers its basic services for a monthly fee. The basic service lets you view jobs and create a profile with your bio (your resume). Your bio is then posted online, and you are available to be contacted by potential employers. They also have a premium service which allows you to apply for any job on their network, connect with recruiters, save searches and access professional advice on resumes, networking, interviewing and more. TheLadders lists opportunities that pay $100,000 or more and claim most employers do not post these high-end jobs on traditional job search sites. If your salary expectation is in the six-figure neighborhood, then you might want to try TheLadders for a month or so. But be aware that while they do target $100K plus positions, they do not guarantee every posting hits at or above that number.

A couple of other PFP sites worth noting: IvyExec.com and RiseSmart.com. I don't recommend you use a PFP job site right away. Try using the free job sites at first and only consider using the paid sites if you feel you are not getting enough opportunities.

Social Networking Sites

I'm going to get more into the subject of Social (Professional) Networking in the next chapter, but wanted to point out here how effective social networking sites can be in discovering opportunities. Take Twitter as an example. Twitter is a social networking and micro-blogging service utilizing instant messaging and has a simple-to-use web interface. Twitter can be a helpful tool when searching for a job. More and more companies are posting open positions on Twitter. Hiring managers can use sites like JobShouts.com to post jobs for free. These jobs are then automatically "tweeted" to users on Twitter and are tagged on several other social networking sites such as LinkedIn and Facebook. Twitter even has its own Job Search engine--TwitterJobSearch.com. Again, I'm going to get more into this in the next chapter, but wanted to mention it here since being digitally searchable is as important as knowing how to search.

What about Craigslist? (www.craigslist.org)

Craigslist is a centralized network of online communities, featuring free online classified advertisements with a whole section devoted to jobs. Craigslist is a good source of job listings in specific locations. Jobs that are listed in your local classifieds typically make it to Craigslist. You can also post your resume to your local Craigslist site. And since Craigslist gets visited often by search engines, its a good idea to have your resume there loaded with relevant key words, especially in your resume title. (One tip, if you decide to post your resume on Craigslist, and you want it to remain confidential, use the anonymous email response and take your name and address off your resume before posting.)

"There's an App for that!"

I certainly can't speak of job search websites without mentioning their mobile phone counterparts. After all, it's important that you are armed with all the tools necessary to help provide you with an advantage in your campaign. There are a growing number of useful iPhone and Smartphone apps available to you for your job search. I'm going to mention five here and describe how they can assist you with your digital Search.

1. CareerBuilder.com iPhone App.
The CareerBuilder iPhone application enables you to search for a job from wherever you are. Using the iPhone's built-in geolocation technology, this application automatically

determines your location and then allows you to find a job using a simple keyword search. The CareerBuilder.com iPhone app was one of the first job search apps built and remains quite popular because of its brand name. It's a free application and best of all has the full functionality of the CareerBuilder.com website. The beauty of using an iPhone to search for a job is that its your private network. Even if you are using your lunch break to job search, its not appropriate to use your current employers network to search job boards. But thanks to the developers at CareerBuilder you can look for jobs in the safety of your cell phone anytime and anywhere. Imagine you are stuck on a crowded train wishing you had a job with an easier commute. Just get on your iPhone and begin looking for that next great position, one that is closer to home. This app can be downloaded from the CareerBuilder.com site.

2. LinkedIn iPhone App.
The LinkedIn iPhone App puts your professional network just a touch away. You can browse through your LinkedIn connections and search for jobs. There is also the ability for you to save your search history and results. Downloading the app is simple and free. You can find it on iTunes and in the iPhone App store. We'll talk more about the LinkedIn website in the next chapter.

3. Job Finder iPhone App.
Job Finder works like SimplyHired in that it aggregates jobs from multiple job sites, saving you from having to search them all individually. With the Job Finder app, all you do is specify the locations you're interested in and set a few preferences—the app does the rest. Job Finder costs about 99¢ and can be downloaded from iTunes.

4. iJobs iPhone App.
The iJobs app delivers millions of job listings to your iPhone. It also has good search features. There is a key word search that allows for searching against a word in a title, description, salary, and more. This app also has an auto detect on your location, a zip code search, and a feature that automatically saves your last search. Similar to the other applications, you can add jobs to your list of favorites and filter the results by category, company, title, and position. This app cost about 99¢ and can be downloaded from iTunes.

5. Craigster iPhone App.
Craigster is, as you might have guessed, the iPhone application for CraigsList. Just like with CraigsList.com you can search or browse through thousands of classifieds, bookmark your favorite searches, email listings to yourself, and call numbers directly from listings. Craigster costs about 99¢ and can be downloaded from iTunes.

Most of these iPhone apps have a Droid and Blackberry equivalent. Simply Google the app name along with your smartphone type to find the appropriate application for you.

Chapter Ten

Develop Your Professional Network

I couldn't wait for success, so I went ahead without it.
—Jonathan Winters

In the last chapter (step 9) I discussed the best sites to visit and the right techniques to use in order to *find that perfect job for you*. In this chapter (step 10), I am going to show you the best sites to visit and the right techniques to use for *that perfect job to find **you***.

In his book on social networking, *Throwing Sheep in the Boardroom*, Matthew Fraser refers to what he calls "kindness of strangers". Fraser is referring to how most people will help others even without knowing them or being introduced to them. Seems it's a natural human tendency to help others in need with the notion that they will help us when we are in need. And that is really what social networking is all about, connecting with others, helping others, and sharing information. That information may lead you to your next great opportunity.

Online social networking was once something mostly left to teenagers looking for the latest gizmo, cool band, or hot date. Not anymore. Social networking sites are a global phenomenon. Sites like MySpace, Facebook, Twitter, and LinkedIn now boast hundreds of millions of members. Online social interaction has become an indispensible part of our daily lives and professional recruiters know it. They are now hunting for job candidates in all the popular social networking sites.

For your job search, online social networking will enable you to push beyond the 10 to 15 best friends that anthropologists say we typically have or even the 150 that we can maintain as steady acquaintances. An online social presence where you can showcase your skills and experience is a must-have during your job search campaign because you don't have to know people intimately to connect with them and to share information with them. Information such as, "You are a person looking for the right opportunity," or "they have an opportunity looking for the right person". Perhaps you don't know that most hiring managers will Google a job candidate at some point in the hiring process—sometimes before an interview. They do it all the time and you would be surprised what interesting tidbits of information shows up in the search results. Google yourself and find out what happens. If you are already in LinkedIn, your profile will most likely rank high in the search results—unless you have an extremely common name.

As we mentioned, this step will take you through the process of establishing (or enhancing) your social networking presence. If you feel you already have a good online presence, this step might feel a bit redundant. And if you are indeed already out there, please check to make sure the content is appropriate for job hunting. In other words, make sure when you Google yourself, you are not presented with photos from your fraternity days overindulging during the homecoming tailgate party. Make sure the content that a potential employer can access is appropriate—clean up the stuff that's not. Just realize that if you are looking for a job, anything posted on the internet is fair game for any employer to see. Online social networking will greatly enhance your job opportunities when used correctly, but when used the wrong way, it can backfire and jeopardize a job offer or even your current job.

Before we get into the key social networking sites available to you at no cost, I have a small exercise for you. It's an exercise to complete your **CoM** list or "**Contact of Mine**". Essentially you will list all the contacts in your current network. Now you may think that you don't have a network, but you do. Or perhaps you think your network is small and of little value to your campaign. Wrong, your network is large. Larger that you realize and full of hot contacts. Contacts that have contacts. Contacts that want to help and will help.

Your **CoM** includes:

- Family, friends of your family, and your neighbors.
- Friends you grew up with, from high school, college, and alumni networks.
- Previous teachers, professors, and coaches.
- Past co-workers, supervisors, or bosses.
- Customers, vendors, and recruiters you have worked with.
- Managers at competing companies who you know.
- Parents from your kid's boy scout/girl scout troops.
- Parents from your kid's little league, soccer teams, lacrosse teams etc.
- People you have met through your Church or Temple.
- Civic organizations in which you may belong (i.e. Rotary, Kiwanis, etc.).
- People you meet at Starbucks or even on an airplane!

Use the sheet at the end of the chapter to write down all the **CoMs** that you can think of. Make additional copies of this sheet as needed. Set a goal of writing down at least 50 contacts.

Once you have your **CoM** list completed, you are ready to begin your online social networking experience. Done right, this process will prove to be the most effective step during your job search campaign. And you'll be happy to know that we are not recommending you email each of your contacts out of the blue to tell them that you are searching for a job and ask them to let you know if they see or hear about an available position. That would prove to be quite awkward and very ineffective. Instead you will use your **CoM** list as your primary resource while you build a strong social networking presence online.

LinkedIn.com

If you are not already on LinkedIn, you definitely need to be. LinkedIn is clearly the number one professional networking site. It is extremely popular, simple to use and is, for the most part, free. LinkedIn connects those with jobs and those looking for work. LinkedIn claims to add more than one million new members every month with close to 50 million signed up worldwide. That's a very large network, and it is all available to YOU, for free! On the LinkedIn web site you can setup a profile showing your current and past positions as well as what you are presently working on. On LinkedIn, you showcase your skills, experience and education. Basically, you write your resume in your LinkedIn profile. Most recruiters use LinkedIn as a place for sourcing candidates because they know all the top professionals are on there.

Build an effective LinkedIn Profile in 3 Simple Steps

Step 1 - Join the Network

Navigate to www.LinkedIn.com. Create a user-ID and password. Follow an easy process of creating your professional profile, including employment (current and past), education, and some other optional information such as memberships and affiliations you have with professional groups and associations. What really makes this so easy is that you already have all this information at your finger tips—the worksheets from steps 4 through 8. Make sure to include all your resume keywords and skills in your profile, so your profile will be found later. If you are looking for a certain position, you can state that in your profile and all your contacts will see that you are indeed in the market for a certain position. They will be able to further drill down into your qualifications, experiences, and skill-set by taking a look at your resume. Keep in mind, LinkedIn is just like any other search engine recruiters are using—keywords are extremely important. Be sure to fill out your entire profile, just like you would a resume.

Step 2 - Build your Network

Now that you are a member, you can connect with other members and build your network. The more connections you have, the more opportunities you will see. At first, connect to people you know and trust or have a business relationship with. To connect, go through your **CoM** list one by one and perform a people-search on each person—the people-search function is on your LinkedIn home page. I know what you are thinking, you have fifty plus folks on your list and therefore you think this is going to be a very long and tedious process. Actually no. It's really going to be quite easy and might even be fun. I bet you haven't heard from most of the people on the list in ages. Think of how interesting it's going to be finding them and discovering what they are currently doing and the things they have accomplished since you last connected! When you find a contact, go through the **"Add Contact to your Network"** process. Attach a brief but personalized note and send the invitations. When the contact receives your invitation, they only have to accept and just like that you are connected. Continue until you have sent invitations to everyone on your **CoM** list that you find through the people-search.

Step 3 - Get Recommendations

As your connection list grows, you should begin to seek recommendations from people you are currently working with or have worked with in the past. Navigate to the "Request Recommendations" function. Send requests to professional associates that you trust will give you a favorable recommendation. Recruiters and hiring managers do read and consider these especially if they are coming from high ranking leaders.

Once you are set on LinkedIn, you will have access to the many powerful tools available to you such as the "Jobs Search" function. LinkedIn also has many different groups that you can join including: college and company alumni groups, community service groups, veterans groups, etc. Within these groups there is a discussion area where you can find open position postings and where you can post your availability for a position.

I should mention that LinkedIn does offer additional services for a fee. After all, they want to be successful too and *we* certainly want them to continue to provide this great service. You can, if you would like, pay for a premium account which will give you access to premium services such as the ability to contact recruiters for help and advice with your job search. For now, I recommend you stick to the free services and see what happens.

Make LinkedIn your **number one** social networking site during your campaign.

Facebook.com

Facebook is one of the world's largest and most popular social networking sites with millions of users around the world. Even though Facebook is not necessarily geared towards professional networking, it can indeed be used as a job hunting tool. It's a great place to keep in touch with friends and family while soliciting them for assistance in a job search. You can also go to the Facebook marketplace, which lists job openings and/or other opportunities in your network. When you find a job opening that you're interested in, you'll be able to message the hiring manager directly. I also recommend joining groups and fan pages to find people with common interests and to network with them. Make Facebook your **number two** social networking site during your job search campaign.

Here are a few simple tips in using Facebook for Professional Networking:

1. Create a simple profile with minimal graphics and widgets.
2. Limit the photos you post.
3. Post content relevant to your job search or career.
4. Choose your Friends (connections) wisely.

Twitter.com

Twitter is a web site for friends, family, and coworkers to communicate and stay connected through the exchange of frequent answers to one simple question: What are you doing? Wikipedia describes Twitter as, "A free social networking and micro-blogging service that allows users to send text-based posts (tweets) up to 140 characters long to the Twitter website." All updates are displayed on the user's profile page and instantly delivered to other users who have signed up to receive them.

Twitter is a great tool to leverage for your job hunt. You can tweet yourself to a job opportunity 140 characters at a time! **It has been done!** Statistics show that job search networking is much more effective when you make "loose" connections—

remember Fraser's reference to the *Kindness of Strangers*. Twitter allows you to connect with people beyond your immediate circle whose networks and contacts are much different from your own.

Keep it simple. Twitter does not have to be a time-consuming part of your job search campaign. Spend a few minutes a day reading and writing posts. Read what other people write and respond. Join conversations and start your own. Make Twitter your **number three** social networking site during your campaign.

Elevator Speech

The technique of crafting an effective "elevator speech" (also referred to as a "30-second sound bite") is a short exercise that will prepare you well for those precious moments during social networking when you have less than a minute (or less than 140 characters) to describe what you bring to the table. Envision yourself walking into an elevator at a large company where you happen to exchange greetings with the CEO or vice president or some other important decision maker. The scenario goes like this: CEO says, "I don't think I recognize you, what do you do around here?" .... You have about 30 seconds to tell him the value that you bring or will bring to his organization. Go!

You will be using this technique in a number of different situations during your campaign, for example, while you are speaking with a potential recruiter, or while you are Tweeting or writing on someone's Facebook wall. Or perhaps you are in a casual conversation with someone you just met at the YMCA during your workout and you are telling them what you do and what you are looking for. There are a number of scenarios where this technique will come in quite handy for you. The key here is to write it down and rehearse ahead of time so that you are well prepared for when you have to give your elevator speech. I mentioned that this is also known as the "30-second sound bite". It doesn't have to be 30 seconds, it could be a little longer, but in no case should your elevator speech exceed one minute.

Your activity here will be to write an outline for your elevator speech and then to rehearse it. The outline will be of the form:

I. Background

II. Achievements

III. Goal

IV. Impact

Here is an example:

I. Background:

I have worked in computer operations for the past 10 years, most recently with Moss-Corp in Orlando, Florida.

II. Achievements:

I led teams that built two new data centers with over 100 virtual servers each and delivered a comprehensive infrastructure support strategy.

III. Goal:

I'm looking for a similar opportunity to be able to use my technical expertise to enhance an organization's IT operations.

IV. Impact:

I am an experienced and seasoned manager, and will bring a level of maturity, insight and wisdom to any organization.

Use the sheet on the next page to write your Elevator Speech. You will need to rehearse the speech so it sounds natural when you deliver it—not programmed or memorized. To that end, you don't necessarily need to write out your speech word for word, but rather, write key words that you can remember so that when you actually speak, it will be in your language of the moment, capturing the key message while sounding natural and authentic.

Final thought: Your passion for technology

During the many opportunities you will have throughout your campaign to interact with potential employers, you must always be ready to demonstrate your passion for technology. I used to work in the timeshare industry. A timeshare company lives and dies by its ability to sell intervals (usually in increments of a week) of vacation property to buyers. Sometimes the customer is "hot", meaning they are ready to purchase now and very little selling (convincing) is necessary. But more often than not, the customer is cold. They need to be sold, they need to be convinced that the salesperson is offering them a chance of a lifetime that cannot be passed up. We have all been offered a "free" three-day weekend trip to a high-end luxury resort at one point or another where your only obligation was to listen to a sales presentations. The majority of folks that accept these "discovery" offers are going only because it's free. They plan to fulfill their obligation by attending the sales tour, but have no intention of purchasing a timeshare unit. Guess what? Many of them happily become timeshare owners. They buy because after spending time with an enthusiastic salesperson, they change their minds. Passion leading to enthusiasm is *contagious*. **You** are the salesperson during your campaign and the product you are selling is ***you***. If you demonstrate passion and enthusiasm during your encounters with potential buyers (employers), you will convince them that you have the right stuff and the opportunity for them to get you onboard cannot be passed up.

Elevator Speech Outline:

I. Background:

___

II. Achievements:

___

III. Goal:

___

IV. Impact: [Remember to focus on the type of impact you can have in a future organization.]

___

Name	Company	Association	Email	Phone	Mobile	LinkedIn	Facebook	Twitter
Bill Stockwell	Candle Box Inc.	DPLL	BillStock@.com	(407) 555-5555	(407) 555-5555	Yes	Yes	WStock123

Chapter Eleven

Sharpen Your Interviewing Techniques

When your work speaks for itself, don't interrupt.
—Henry J. Kaiser

It may seem obvious but the best way to ace your next job interview is to prepare for it. As an IT leader, a large portion of my time is spent interviewing potential candidates for both contract and full-time IT positions. I interview for positions that report directly to me and I also assist peer managers when they are interviewing candidates. It never ceases to amaze me how many prospective employees come unprepared and have little or no knowledge of the company and of the company's industry. Some candidates don't even know the specifics of the open position. Don't be one of these misguided souls! Do some homework, get prepared, and give yourself the best possible opportunity.

The face-to-face interview is most often the third stage of the interview process. Your potential employer has reviewed your resume (stage 1), has spoken to you on the phone (stage 2), and now wants to bring you in for a face-to-face interview. They like what they see so far and they want to know more. Of course, the same could be said for the other dozen candidates competing for the same position. However, you have an edge. You have an edge because you read this guide and, specifically this chapter which will show you techniques that will maximize the effectiveness of your interview. In this chapter (step 11), I will show you what is important and not important in an interview and how to answer questions using the **SAI**s that you previously developed in step 6. I will also describe the behavioral interviewing technique which is becoming more and more popular now among corporations and hiring managers.

Always keep in mind that the best qualified candidate is not necessarily the one who gets the position. It is typically the candidate who is most compatible with the culture and who can also effectively do the job and bring value to the company. Notice that word "compatible" which is a very subjective quality. I will provide you with practical methods that you can use to demonstrate your compatibility with the hiring manager and the company.

I had one candidate apply for a programmer position who was very well qualified in terms of his skills. My company was always looking for people who had a particular niche programmer skill set because there were so few of these folks out in the marketplace. So any time I found someone with the appropriate skills who was looking for a position, I would fly them down and invite them to sit for a set of interviews. Our company's interview process included a skills test to verify the candidate's knowledge and understanding of this specialized programming language. One particular candidate I had found did quite well on the skills test and was obviously an experienced programmer. However, during the interview, he came across as abrasive and arrogant. Needless to say, I passed on this candidate and kept looking.

The point here is that managers hire people they like and people hire people they know and can relate with. The most competent and qualified individual does not necessarily get the job, but rather the one who is most compatible (along with being qualified) will be the one who gets the job. Think about this from the point of view of the hiring manager. They are making a decision which could either allow them to meet and exceed their work objectives, or it could be a disastrous nightmare of team dysfunction resulting in project delays or complete failure. A lot is riding on this decision for the hiring manager. Your job as the candidate is to make them feel comfortable that you are a team player willing to go the extra mile to enable the organization to achieve success.

There are actions that you can take during the interview that will enhance your ability to relate to the interviewers:

1) Project a positive image. This means dressing professionally, and speaking in a positive manner about your past employers and past experiences. A lot of this comes down to non-verbal communications as well. Smile a lot and communicate through your expressions and actions that you are excited about the opportunity you have to work for their company and you can't wait to go to work for them because it is an opportunity of a lifetime!

2) Be familiar with the company and the interviewer. This should go without saying. There is plenty of opportunity to research and understand what the company does by going to their website. Make sure you also check out any recent news especially as it relates to company growth. Try looking up the interviewer in LinkedIn, you'll get a good flavor for his/her background which might help you anticipate the types of questions you'll be asked.

3) Pay a compliment. This may come out of #2 above. If you like the company's web site design, make note of that at the appropriate time. This can be especially powerful if you know ahead of time that the person interviewing you helped do the web design (or whatever it is that you are complimenting). Don't just pay a phony compliment, but rather comment on something—some aspect of the business or the company—that you are truly impressed with. That will ensure you sound genuine. One note of caution: stay away from complimenting or commenting on the interviewer's family pictures, trophies they may have in their office, or similar personal items. Some people have these items in their office for their own enjoyment and don't really like others commenting on them unless they are close friends.

There is probably no way that I can emphasize enough how important it is to *practice interviewing*. The more you practice, the more likely you are to have a successful interview. Practice the interview flow as outlined in this chapter and get feedback. You will need a coach, spouse, friend or parent to help you here, but the investment of time into practicing your interview is well worth it. There are very few moments in your life where much of the rest of your life can have such an impact as those 30-45 minutes of an interview. Doesn't it warrant a few practice sessions? Before a lawyer goes into court, he practices in front of other associates or even mock juries. Sometimes the case warrants such large expense because it is so important to the life of the client. Your interview also deserves such care and diligence.

Preparation

I've been coaching little league baseball for years. And believe me when I tell you that 9 year-olds hate to practice, they just want to play. I find myself constantly telling them that the degree of success during game day is directly proportionate to the amount of practice and preparation ahead of time. An interview should be approached in the same way. Here is a sample preparation checklist that you should go through for each interview.

1) Get a job description ahead of time. This may already have been provided but if not, call the recruiter or company's HR rep and ask if you can have it. Then read it carefully.

2) Review your set of **SAI**s and match four of them to the qualification or experience criteria as listed in the job description. If one of the items in the job description reminds you of a good **SAI** that you can write, by all means go ahead and write up a new **SAI**. Your stories that demonstrate a Situation, Action and Impact will be a powerful tool for you to use to communicate your skills and experience during the interview.

3) Verbally rehearse your selected **SAI**s. It is even better if you can practice with a spouse, friend or coach so you can get some honest feedback. Don't feel like you have to memorize word for word what you wrote down because you want to sound natural. The good thing is that you have actually lived these stories so just by recalling the key elements in each section, you should be able to recount them pretty easily and naturally.

4) Spend an hour researching the company, particularly news items. During your interview, you will be given the opportunity to ask questions and it will be impressive if you demonstrate you have done your homework by asking a relevant, thoughtful, and intelligent question regarding the company. If you know someone who already works at the company, this is even better because you can solicit some intelligence from them.

5) Be presentable. Even in the civilian world, there are uniforms for interviews: suits for men and the appropriate conservative outfits for women. Don't wait until the day of the interview to go searching for your suit only to find it miserably wrinkled up on the floor of your closet. Ideally, ensure that you have a cleaned and pressed suit or appropriate outfit ready to go at least the day before the interview. Don't forget a nice pair of shined shoes.

6) Know where you are going and plan to arrive 15 minutes early. I had one candidate arrive for an interview 2 hours early. I don't suggest that because it is awkward to be sitting outside someone's office for a couple hours. However, it is a good idea to know where you are going ahead of time (perhaps drive by the office the day before) and plan to arrive early. I once had an interview in Chicago and I thought I knew how to get there but as I was driving to the site, I ran into a street that was unexpectedly closed and I had to quickly find a detour. Fortunately, I made it for the interview in the nick of time because I had given myself a 20 minute margin.

7) Bring several copies of your resume. Most likely, you will be interviewing with more than one person.

8) Bring a notepad and a pen and take notes, it shows you came prepared.

9) Bring sample work product. Demonstrating samples of your work is a no-brainer in some fields such as graphic design, but it can be even more powerful in other fields that you might not intuitively think about. For example, I used to interview business analysts

and whenever a candidate came in with a sample software development requirements document that they drafted, I was usually very impressed (of course, the documents were sanitized to exclude any proprietary information). Think about ways to show the high quality work that you produce. For example: All you Web designers and software developers, you already have a lot of ways to demonstrate the quality of your work by mentioning websites your worked on and pointing out the actual URLs. We had one delivery manager who brought in a presentation demonstrating how he was able to get his previous company to adopt a new software development methodology through his carefully presented plan. This was a very effective demonstration of the exact skills that the company was looking for in a manager and the candidate was immediately offered the position.

Interview flow

Here is a typical flow of the interview and suggestions on how you can respond and influence it.

Introduction. You walk into the room, greet the interviewer (or interviewers) shake hands with everyone, and you are invited to take a seat. There will be some small talk about your travel or about the weather to help everyone relax. The interviewer may offer you coffee or a cold drink. I find it best to say "no thank you" on the drink. The last thing you need is to spill hot coffee on your lap while you are speaking.

Once you are sitting, ask if you may take notes. This will show the interviewer that you value their time and are taking this session very seriously and are there as a possible solution for them.

Interviewer: "So tell me about your experience at XYZCorp."

Candidate: "I've been with XYZCorp for several years. Started there as a project coordinator and worked my way up to project manager and eventually to program manager. What area are you interested in hearing about?"

Interviewer: "Well, we are looking for a project manager, so why don't you tell me about your experience in that area."

Candidate: "Sure, I'd be happy to. I had been asked to take on the PM role for a systems integration project which had fallen seriously behind schedule and was in danger of getting cancelled unless it could be turned around. I called a meeting of the development managers and project leads and determined that part of the problem was changing requirements. I met with the customer and discussed with them the seriousness of the situation and the need to put in place a change control procedure. She agreed to this procedure and to abide by it which resulted in renewed focus on the work by the development team which led to the project being completed within its revised schedule. The customer was so happy with the outcome that she steered additional business to our unit rather than to our competitors. Is this the kind of PM experience that you are searching for, Mr. Smith?"

Notice the strategic insertion of an **SAI** into answering the interviewer's question. Describing the situation you found yourself in, then how you turned it around (the Action) and the final impact of that action is a very effective way to answer the question. The interviewer will see that you are credible and, many times by the way, the interviewer will be able to relate to you because they have gone through similar trials and tribulations in their own career. There is a good chance that they will empathize with you being caught in a difficult situation and applaud you for overcoming it. They will think, "Boy, I need another person like me in this job working for me to get the job done." Again, don't underestimate the power of a good story—or in our case, the power of a good **SAI.**

Another item to point out is notice the "control question" at the end of the candidate's response. By asking this question, the candidate will get a good understanding of whether or not they are on the right track with the interviewer, or if they need to adjust their answers accordingly. For example, if the interviewer responded by saying, "Well, no, I was looking more in the area of project estimating," you can then adjust by responding with an appropriate **SAI** in the area of project estimating.

Let's look at some other *typical* questions asked during an interview and strategies that you can formulate ahead of time on how to respond.

Interviewer: "Can you tell me what happened at XYZCorp and why you are no longer there?" Or a variation: "Why are you here – why are you looking to leave XYZCorp?" Another, more direct variation: "What did you not like about one of your previous companies?"

Candidate: "When I joined XYZCorp, it was a dynamic and growing company but because of the economy, things have stagnated and I no longer feel like I can make an effective contribution. The folks at XYZCorp are great and I am very proud of my time there and what we were able to accomplish. Now I would like to take on new challenges."

You may have had a difficult time at your previous employer—perhaps you were laid off or were fired. The key point to remember, no matter what the circumstances, is not to dwell on the nitty gritty negative details of your prior situation, but rather highlight the good stuff and how you helped to make your prior employer successful. Never ever speak ill will of your boss or former boss. Never mention any power struggles or politics or anything of the sort since doing so would only reflect poorly upon you. By the way, interviewers will sometimes use such subtle questions as a test to see if the candidate will bad-mouth their previous employer. Their reasoning is that if they will bad-mouth the previous guy, they will not hesitate to bad-mouth their new boss.

In the movie "Liar, Liar," actor Jim Carrey portrayed a lawyer who shared his honest feelings to all around him. Sharing honest feelings of anger, frustration, and hate, is never appropriate during interviews. Don't go there.

Interviewer: "What are your strengths?" (refer back to step 4)

Candidate: "One of my strengths is the ability to figure out new ways to solve old problems and then implement a solution. For example.... [insert **SAI** here]"

This is a common question and is typically asked because the interviewer is trying to a) understand more about you and, b) gauge your level of self confidence. Your answer should be confident, but not cocky. Use of an **SAI** here is perfect especially if you tailor your **SAI** to something related to the requirements of the job you are applying for. Preparing ahead of time to answer this specific question prior to each interview will be very helpful.

Interviewer: "What are your weaknesses?"

Candidate: "I can't sit still. I don't like to just waste time. I need to always be engaged and doing something productive."

This is also another common question and the key here is to never reveal a real liability (like, "Well I'm not usually concerned with over testing any product before it goes live, I'd rather hit the deadline than to ensure the code is 100% functional." Probably not a good idea to reveal during a job interview.) Also try not to respond to the weakness question with a strength (like, "I care too much and work too hard. I stay late and even come in on the weekends"). The interviewer will not appreciate this response, we all have weaknesses. Have one that's not too alarming ready to present. Never ever say, "Oh, well, I don't have any weaknesses."

You should have a stock answer ready to quickly respond to this question. It never fails that when I have asked this question of a candidate during an interview, they have always been caught off guard and, after some squirming, have forced out an answer or simply avoided the question. In some cases, they have revealed a real liability that caused me to pass on their candidacy.

Interviewer: "What are your short and long range goals? Where do you see yourself one year from now? Five years from now?"

Candidate: "I have always been excited and passionate about developing and implementing new technology solutions to help people work better and more effectively. I would like to further advance in my career so that I can have a greater impact on implementing technology solutions—further developing my skills and getting a chance to employ them as a technology delivery leader."

The interviewer wants to understand your thought processes and any career planning that you have done. Typically, they want to hire you to be a successful long term employee. That is good for you, but, more importantly (to them) it is good for the company and it will make the hiring manager look like a genius (and by the way, managers are evaluated based on the quality of the talent that they hire).

Some additional common questions - try to have answers prepared.

What did you like best\least about your last position?
What is the biggest mistake you ever made?
What kind of work environment suits you best?
What did you like\dislike about your last supervisor?

and my favorite - from one of my previous supervisors...
    What 5 adjectives would your last boss use to describe you?

Time now for your questions

Typically towards the end of the interview, the hiring manager will give you the opportunity to ask any questions. You should have one or two prepared ahead of time. Feel free to write them down in the notebook that you brought into the interview with you, and refer to it to refresh your memory and take notes on the responses.

Questions might be of this nature:

-- "What SDLC do you use here, are you exploring other options?"

-- "What is the company's outsourcing strategy, onshore/offshore?"

-- "How is this department organized? Is it all in one location, this one?"

-- "How are the teams structured? Is this more of a matrix organization?"

-- "How would you describe the culture here?"

-- "What are your expectations for me during the next 6 months?"

In fact, that last question may be the best one you can ask and, if you can work it into the interview early enough, it may focus your later answers by conveying that you are able to meet and exceed their expectations.

Note this is not the time to ask about salary and benefits. Defer those questions as far out as possible. The next chapter (step 12) will present an effective strategy that will enable you to discuss this issue from as strong a position as possible.

Assuming the interviewer has given you permission, feel free to take notes on responses to your questions. Remember you are also evaluating THEM to see if they are a fit for what you are looking for in an employer in your search for the perfect job.

Feedback question

There are a couple subtle ways you can determine the reaction of the interviewer to your candidacy for the position offered. One way to do this is to ask a feedback question such as: "How do you see me as a fit within the company?" -- or -- "Do you feel I would make a good fit within XYZCorp?"

If you get a positive response, you should follow up with, "That sounds great! Where do we go from here?" -- or – "That sounds great, what is our next step?"

If you get a somewhat negative or ambivalent response to your feedback question, you should try and determine the basis for that. (i.e. whether your skill set wasn't a match, or you did not have enough experience in a certain area, etc). Use follow-up questions so that you have a clearer understanding of where you stand and if you feel they don't have the full picture and you do indeed have experience in a certain area, this is YOUR CHANCE to jump in and properly inform them – preferably with a good **SAI**!

The Close

Finally, don't let them just say, "We'll get back to you." If they try and say that, ask them when you should expect to hear back from them. If they say, "Oh by next Thursday," make a note of that and ask them if you can contact THEM if you do not hear from them by Thursday.

Conclude the interview by standing up, shaking their hand and saying, "Thank you, Mrs. Smith, for your time and consideration, I enjoyed our conversation. Sounds like you have a lot of exciting challenges that I hope to be a part of. I look forward to talking to you on [such-and-such day]."

Post interview

Right after the interview there are a few things you want to do that will be very helpful in your campaign. First and foremost is write a thank you letter to all those who took the time to interview you. This letter can be sent via email - cc the HR manager. Send your email no earlier than 24 hours after the interview, but no later that 48 hours. It will show that you are a class act and will certainly enhance your candidacy.

Treat every interview as a learning opportunity. Write down all the questions that were asked of you and some notes on your responses including the **SAI**s that you used. Make notes on any initial impressions that you had. If you can critique yourself to help you answer such questions in future interviews, write that down as well.

Finally, set a reminder in your calendar on your agreed follow-up contact. If you reached agreement that you would hear back from them by such-and-such date, note that in your calendar, and if you don't hear back, give them a call. Being persistent is never a reason for someone not getting a job, so... be persistent but don't be desperate—you'll know the difference.

Behavioral Interviewing

Many organizations are now practicing a technique for interviewing candidates called "behavioral interviewing". In order to effectively employ this technique, they have trained their managers in how to conduct an interview using these methods. The idea behind this technique is to elicit from a candidate how they have behaved and conducted themselves in certain situations. Behavioral interviewing can reveal much more about a candidate's work habits and interactions with other people because, through its style of questioning, it subtly reveals a candidate's attitudes and motivations.

This is how it works. The interview will typically begin with a statement that the interviewer will look for specific instances from real situations in the candidate's past experiences in response to the interviewer's questions. A typical question will start with the words, "Describe to me a specific situation in which you…" For example, "Describe to me a specific situation in which you had a conflict with a co-worker. What did you do when you knew you were right?" If the answer that the interviewer gets involves some sort of intimidation or threats on the part of the candidate, then obviously, that is very telling.

Another common question is, "Tell me about a specific situation where you became aware of co-worker acting in an unethical way." Personally, I have not had the displeasure of this happening to me. I was however, asked this specific question once during an interview. I responded by stating that the situation had never presented itself. The interviewer quickly followed up with a "what if" question. "What if I had become aware of a co-worker using a company PC in an unethical way? How would I handle it?" My response and the correct response is simply, "I would contact human resources and report the issue. I would also inform my supervisor of the incident and then return to my business. I would not, under any circumstances, tell other employees of the incident." The key in your answer is to demonstrate that you realize you have an obligation to take action and that you know the appropriate process to follow.

Behavioral interviewing works well for hiring companies because it enables them to peel back the veneer of the candidate who is otherwise putting their best foot forward and look for issues that might be a cause for concern. Each company has a culture, and behavioral interviewing helps them understand if the candidate will be a fit or be a source of conflict. How can you succeed if a company uses this technique? Very easily, because the **SAI**s that you have prepared are tailor made for responding to such questions. Your **SAI**s are real stories from your past experiences where you have dealt with conflict and team issues and have succeeded in resolving those in order to meet an objective. These stories should indeed highlight how you were able to effectively work through conflict and collaborate as a team to accomplish your goals.

Final thought: Impact of failure on the road to success

On the way to success you must be ready and able to handle some failure. As I mentioned before, it is not always the best candidate that gets the job. So many factors are in play here, some of which you have no control over. Following the steps that we have outlined in this book will give you the best possible opportunity and will certainly put you ahead of the competition, but even then you might still experience failure. Successful people understand that failure is part of the journey to success. Successful people accept failure, regroup, and formulate a better way to succeed. In most cases, successful people fail many times before they ultimately achieve their goal(s). In his book *Winning the NFL Way*, Bob LaMonte writes about this very concept using Thomas Edison as the example:

> "While inventing the electric light, Edison made 25,000 attempts before he got it right. A reporter once asked him, 'Mr. Edison, how did you deal with 25,000 failures?' Edison replied, 'I did not fail 25,000 times. I was successful in finding 25,000 ways the light bulb didn't work.' Edison had tenacity. He refused to quit. He would not accept failure."

You may have several interviews before getting a job offer. After each interview, write down the questions that you were asked and the **SAI**s that you used as responses. If you had a question where you could not readily produce an **SAI**, go back through your experiences and draft a new **SAI** for the next time this or a similar question comes up. Practice your response. Learn from each interview in order to improve for the next one. By going through this self-critique each time, you are collecting chips in your favor for future success—but it takes the discipline of applying these principles in order to learn and improve.

Chapter Twelve

Seal the Deal

In theory there is no difference between theory and practice. In practice there is.
—Yogi Berra

Dealing with the salary question might be the most difficult part of the interview that you will face. Most of us feel anxious, embarrassed and uncomfortable when the time comes to discuss salaries. Again, the key here is to be prepared. Prior to the interview, research and be aware of the potential salary for the position. You will also want to think about the minimum amount you need in order to support yourself/your family and maintain your lifestyle. As I mentioned in chapter one, think of it like bidding on a house, go to the negotiating table with three numbers in mind, a high, a low and a mid-point. The high is your target and the low is the minimum you'll accept. Knowing your bottom line is important. So decide, before you go into an interview, what salary you want to earn, what you need to live on, and what you will be willing to settle for.

This chapter, your final step, will give you some strategies that you can use to negotiate a favorable salary and benefits package. With every position, there is always room for negotiation. If you practice and employ the methods mentioned here, you will most likely be able to get a better package than you would otherwise. Educating yourself and practicing these techniques can have a tremendous payoff. Think about the issue of compounding. Let's say you learned how to negotiate and you were able to get a salary that is $3,000 higher than you otherwise would have received. That base salary that you just negotiated will stay with you for entire time that you are with that company so let's do a little math and look at what this means. An extra $3,000 over 5 years means $15,000 of extra income that you received. Think you can use an extra $15K? But, wait, it gets even better! Your annual raises are typically tied to a percentage of your base salary. Let's say you get a 4% annual raise. $3,000 multiplied by 4% is an extra $120 per year. Times 5 years is $600 – and that is a straight line calculation to make things simple (it doesn't take into effect the wonders of compounding). Are you eligible for a bonus? If so, then this logic applies here as well because a bonus is also typically a percentage of your annual salary. So... if you get a 12% bonus, that extra $3,000 in salary is $360. Multiplied by 5 years results in an extra $1,800. In the scenario that I have given, you can see that understanding how to negotiate can result in you having an extra $17,400 over the 5 years with that company. Think it is worth it to negotiate effectively?

Here are a few important rules to follow when it comes to salary negotiating.

Rules of Engagement

Rule Number 1

The number one rule of salary negotiation is to try and put off mentioning any number for as long as possible. The longer you can delay any mention of a figure, the better the outcome—for you! Why is this? Because this allows you to negotiate in a position of strength. Think about it. You are coming into the job interview initially in a position of weakness. You like this company and you want the job. In some cases you really want the job. In other cases, heck, you **need** this job! However, as the company's job search progresses and as they get to know you better and like what they see, the power equation starts to shift. The hiring manager wants what you can do for them. They see the value that you represent and the potential that you have to help make THEM successful. Never again will you be in such a position to negotiate so effectively. Seize that moment. Of course, this position will only get better the longer you can hold off on talking about salary.

Rule Number 2

The second primary rule to remember is, "He who speaks first, loses". Whoever mentions a number first has shown their hand and thus starts the bidding. I saw this first hand on one of my job searches. I had put off any mention of a number for as long as possible and then it came time for an offer to be made. My future employer wanted me. I liked what I saw and really wanted to work there. I came in with a number in my head. It was a fairly low number because I was currently out of work and was really needing a job. However, I held my ground and waited until the hiring manager mentioned a number. The number was several thousand dollars higher than the number I went in with. But I didn't show my hand and responded with an even higher number than the hiring manager mentioned. We agreed to then meet in the middle and he was happy and, needless to say, I was very satisfied.

When asked about desired salary, the best response is one that returns the question to the hiring manager. Say something like, "What kind of salary range are you working with?" or, "Well, I'd like to be compensated like other employees in similar positions and with similar qualifications." Or, "What is the typical salary for this position?" Another strategy is to avoid a specific salary and name a pay range instead. Say: "I was thinking of a salary in the $75,000 to $95,000 range."

Rule Number 3
The third rule is to avoid disclosing present (or past) salary. Your negotiating edge goes out the window once your past salary is on the table. You want to force the hiring manager to make his/her best offer by not disclosing exactly what your current salary is or exactly what it would take to get you to leave your current job.

The Process

Early Stage # 1: Defer the issue

Many companies will try and determine what your salary range is very early in their hiring process because they want to screen out candidates with unrealistic compensation expectations. I personally think that is a mistake on their part, because they should want to hire the best candidate first rather than risk "screening them out." But in any case, they attempt to have you screen yourself out. If asked about your "salary requirements" or your "salary history", you should respond in a positive manner. Here are some example responses:

"Salary is a subject that is certainly important and I am sure we can work that out, but before we go there, I think it would be helpful if we talked about the requirements and responsibilities for this particular position. I would like to know more so I can figure out how I can help."

-- or --

"Money is an issue, but I am more interested in making sure that I find the right fit, where my skills and experiences can really make a difference and a positive impact on the company. Can you tell me more about what you are looking for…?"

Middle Stage # 2: Let the games begin!

If they press the issue further or you get the sense that they really want to address salary, then you can use some of the responses below. A note of caution here since, at this stage, it can be very easy to jeopardize the good rapport that you have been able to build up so far. The key is to keep your responses focused on them and not necessarily on what you want. Your response should be similar in tone to the above: you are looking for a good fit, you want to make sure you can make a contribution and have a positive impact, you want everyone to be happy, and then ask THEM for a range. Remember, you want to defer as long as possible, but if someone needs to say a number, you want them to do it first.

"Well, you know I am very interested in developing a relationship with the right organization where I can work with great people and make a positive contribution. Now that you have raised the issue of salary, what is the number you have in mind?"

-- or --

"Certainly salary is an important issue, but I don't have any magic number. What is the range for this position?"

-- or --

"From what we discussed, it looks like the position is a great fit and I am sure I can really make a contribution. What's the salary you have in mind or the range for the position?"

After making your statement and question, be quiet. Do not say a word. Even if you have to sit staring at each other for the next 5 minutes, do not say anything. Let them speak. Hopefully it will be with a number. If it is a number, go straight to to "Late Stage # 4: Final Negotiation".

Middle Stage # 3: Getting down to nuts and bolts.

Let's say the hiring manager is in that 10% of managers who know this game as well as you and they refuse to be the first to divulge a number. They continue to persist by asking you, "How much money do you want?" You should then open up the discussion further. Your goal here is to find out if they are offering you the position, and if so, you can then come to agreement on a salary.

For example: *"Compensation is certainly an important issue, Mr. Smith. Would I be correct in assuming that if we agree here, you are offering me the position? If so, I am sure we can come together on salary."*

Another response: *"Mr. Smith, what is your idea for a fair compensation package for this position?"* This response could open the discussion to bonuses, stock options, vacation time, etc. Any information you glean here would be useful in your final negotiation.

A different variation of their question could be: *"What is your current salary?"* You have a couple options here that may avoid either disqualifying you from the position because your current salary is too high, or it could help you negotiate a better salary:

"My total compensation package has been in the range $X to $Y depending on various factors."

-- or --

"From my understanding of the market and various surveys, the typical salary ranges from $X to $Y."

-- or --

This response you can use if it applies to your circumstance:
"Similar offers I have recently had are in the $Z range."

Late Stage # 4: Final negotiation.

Let's say they were the first to blurt out a number. How do you respond? You should quickly realize whether or not that number meets, exceeds or does not meet your expectations. If it meets or exceeds your expectations, you then need to understand whether or not they are offering you the position before you respond any further. If they have made an offer or are in the process of making an offer, you can make a counter offer that is higher than the number they gave you. Typically, they will have some room for this negotiation and an agreement can be reached. If they have no room on the base salary, you can certainly negotiate the bonus potential. Another thing that is negotiable is asking for a six month review that is tied to your compensation with the expectation that you would get a certain percentage increase at the six month point.

If the number they give is below your expectations and they are offering you the position, then you can certainly negotiate with a counter offer that is at least in line with your expectations or with what you are willing to accept. If you come to an agreement on a salary that is still below your expectations and you have also discussed bonuses, accelerated review schedule, vacation time, etc. and you are still not sure about the salary and the job, then this is when you need to seriously consider your situation and if, given the totality of your circumstances, the position, and the opportunity presented, if this is the right job for you.

Do not make any decision on the spot. There is no reason you need to accept the offer right then and there. Ask for some time to run it by your spouse or significant other. Most companies will gladly give you time to think about the offer. However, there is typically a time limit when they will want a response back from you – usually within one or two days.

Make sure you have a clear understanding of the following before you accept an offer:

- Base salary
- Annual bonus
- Compensation incentives
- Benefit package such as medical/dental (and when those would start)
- Retirement package such as company stock, 401K, and company match

And if applicable:

- Relocation package
- Signing bonus
- Company perks

Also make sure you are aware of the start-date and next review-date. In fact, it is best to get all the above in writing before you make your final decision. This will be one of the most important decisions in your life so it makes sense to approach it will all the due diligence and care that you would any momentous decision.

Just to make one more point regarding salary negotiation. It is important to role play this aspect of the job search as well because this can be a tricky subject for you emotionally. You don't want to come across as so eager and hungry that you appear desperate. Even if you are, don't appear that way because you will not be doing yourself justice or your family any favors by taking a lower offer than you deserve. At the same time, you don't want to give up a good position with a lot of great potential because you were not flexible enough on salary. Practice these techniques with someone acting as the hiring manager and get feedback on your tone and how you present yourself. You should maintain a positive tone where you demonstrate that you are excited about the opportunity, yet you are confident in your ability and worth to the company and expect to be compensated fairly.

Additional Points

Don't discount the value of the non-salary benefits of the compensation package. I'm taking about medical benefits, stock offerings, company perks etc. Sometimes the salary offered may seem low, low enough for you to turn down the job. But benefits and perks can add up to more than a third of the overall compensation. Some benefits are fixed such as medical plans, but others are negotiable such as stock options, bonuses, employee discounts, training, flexible working hours, personal time off and sick time.

Always be amicable when discussing salary with your potential employer. You should help them understand that you are on the same side and working together to find a package that would satisfy both parties. Be sure to make your salary discussion a friendly experience, treat every offer seriously and graciously even if you decide not to accept the offer. And if you decline, be sure to leave on the best of terms. You never know when other opportunities may arise with this same employer, so don't burn any bridges.

The popular sitcom Seinfeld is among my favorites of all time. I think each and every show is pure comedic genius. On one particular show the character Kramer is suing a coffee shop company because he spilled coffee, coffee he claimed was too hot, on his lap, and was burned. Prior to the settlement meeting, the legal team representing the coffee company agreed to offer Kramer a lifetime of coffee and a large monetary settlement. When Kramer and his lawyer, Jackie Chiles entered the room, the legal team began to verbalize their offering. *"We are prepared to offer you a lifetime of free coffee at all of our shops worldwide and ....."* and before they had a chance to add the monetary settlement Kramer jumped up out of his chair and blurted ***"I'll TAKE IT!"*** Needless to say they shook hands and Kramer walked away with lots of coffee but with no money.

My point here is to hold off on saying yes to an offer right away. Be enthusiastic and appreciative when you get the job offer, but ask for some time to think it over. This gives you time to get over your initial joy at being selected. Go home, think it over, sleep on it, consider all the aspects, and then go ahead and say ***'Yes, I accept'***. Then open a bottle of champagne and celebrate the beginning of the next chapter in your career working for a great company, doing the work you love, being properly compensated for your skills, and enjoying the job you've been dreaming about your whole life.

Congratulations, you have completed the 12 steps and now you are ready to find your dream job. Good luck to you, you are going to be great!

BIBLIOGRAPHY

The following books were mentioned in this book and are recommended by the authors for further professional reading.

Bob LaMonte, *Winning the NFL Way*. Haper Paperbacks. New York, 2005.

Joe Torre. *The Yankee Years*. Doubleday. New York, 2009.

Marcus Buckingham, Donald O. Clifton. *Now, Discover Your Strengths*. The Free Press. New York, 2001.

Mark Sanborn. *You Don't Need a Title to be a Leader: How Anyone, Anywhere, Can Make a Positive Difference.* Broadway Business. New York, 2006.

Matthew Fraser. *Throwing Sheep in the Boardroom: How Online Social Networking Will Transform Your Life, Work, and World.* John Wiley & Sons. Hoboken, NJ, 2008.

Stephen R. Covey. *The 7 Habits of Highly Effective People*. The Free Press. New York, 2004.

Stephen Denning. *Squirrel, Inc.: A Fable of Leadership through Storytelling*. John Wiley & Sons. San Francisco, 2004.

Acknowledgements

The authors would like to acknowledge and thank those who have contributed to this book with your, comments, suggestions and guidance: Urmila McClure, Gina O'Sullivan, Katrina McClure, Dalia Rojas, Bob Hiltz, Debra Johnson, Julie Bockover, Terry Loewen, Larry Mulvaney, Kelly Lambert, and Greg Rose. Thank you!

About the Authors

Ian O'Sullivan has more than twenty years experience in the Information Technology field including extensive management and leadership positions with Fortune 500 companies in the areas of application development and solution delivery. As a Technology Management Consultant, Ian implemented software construction best practices optimizing product quality, reliability and timeliness for major U.S. corporations. Ian is an expert at building and leading successful teams in delivering enterprise level software solutions. He currently resides in Orlando, Florida with his wife and two sons.

John McClure is president of Signalman Publishing which specializes in bringing non-fiction works to e-readers everywhere. He has a varied career history including time spent as a nuclear trained submarine officer in the U.S. Navy and twelve years experience in corporate America as an Information Technology manager. He currently resides in the Orlando, Florida area with his wife and daughter.

QUICK ORDER FORM

Email orders: orders@signalmanpublishing.com

Postal orders:
Signalman Publishing
3209 Stonehurst Cir
Kissimmee, FL 34741

Please send the following books. I understand that I may return them for a full refund, for any reason--no questions asked.

_____

_____

Indicate quantity above.

Name: _____

Address: _____

City: _____ State:_____ Zip:_____

Country:_____

Email: _____

Sales tax: Please add 7% for products shipped to a Florida address.

Shipping by air US: $4 for first book and $1 for each additional.
International: $9 for first book and $4 for each additional.

QUICK ORDER FORM

Email orders: orders@signalmanpublishing.com

Postal orders:
Signalman Publishing
3209 Stonehurst Cir
Kissimmee, FL 34741

Please send the following books. I understand that I may return them for a full refund, for any reason--no questions asked.

_____

_____

Indicate quantity above.

Name: _____

Address: _____

City: _____ State:_____ Zip:_____

Country:_____

Email: _____

Sales tax: Please add 7% for products shipped to a Florida address.

Shipping by air US: $4 for first book and $1 for each additional.
International: $9 for first book and $4 for each additional.

www.ingramcontent.com/pod-product-compliance
Lightning Source LLC
Chambersburg PA
CBHW080443110426
42743CB00016B/3264